THE PHYSICS
OF EINSTEIN

Black holes, time travel, distant starlight, E=mc²

JASON LISLE

First Printing: January 2018

Biblical Science Institute, P.O. Box 1414, Aledo, TX 76008

ISBN: 978-0-9997079-0-6

Cover design by Chafi Charneco (www.chafish.com)

Printed in the United States of America

Please visit our website for other fascinating articles and resources:

www.biblicalscienceinstitute.com

Acknowledgements

I deeply appreciate the support and encouragement of my family and friends and of the supporters and partners of the Biblical Science Institute. I especially want to thank Dr. Jake Hebert and Dr. Vernon Cupps for reviewing this book and providing helpful feedback and I thank Terri Kammerzell for editing. I pray the book stimulates interest in physics and a deeper love for the God of creation.

Contents

Introduction

Is it possible to travel faster than the speed of light? Will future human beings build spaceships that can travel at "warp" speed like in *Star Trek*? Is time travel possible? If so, could we ever travel back in time to prevent a catastrophe from occurring? What does $E = mc^2$ really mean? What are black holes, and do they really exist? What would happen to a person who fell into a black hole, and how do we know? Is the universe really expanding? How long does it take starlight to travel from distant galaxies to Earth? Does this distant starlight require the universe to be billions of years old?

The branch of physics pioneered by Albert Einstein allows us to answer all the above questions. And the answers are not what most people would expect. The universe turns out to be far more interesting than human beings could have ever imagined. The laws that govern space and time are wonderful and possess an elegant complexity indicative of the creativity of God. The physics of Einstein gives us a finite peek into an infinite mind, allowing us to answer questions pertaining to time travel, black holes, and the nature of light. And it does so in a way that is remarkably conclusive, far more so than is generally possible in other branches of science. We begin our exploration of this topic with some historical background.

Albert Einstein (1879-1955) was a German-born physicist who made a number of contributions to our understanding of the universe. Einstein lived in Switzerland for a time, and in 1933 moved to the United States where he became a citizen in 1940. He was Jewish by descent, but not by religious conviction. He apparently believed in a god, but in a deistic or pantheistic sense – not the personal, living God of Scripture. Nonetheless, his research reveals the handiwork of God.

Einstein made a number of significant discoveries in physics. These include the discovery of the photoelectric effect in 1905, for which he won the Nobel Prize in Physics in 1922. His most significant discoveries, and those for which he is most well-known, involve the branch of physics often referred to as "the theory of relativity." This fascinating area of science deals with the nature of space and time, and the counterintuitive effects that happen as an object approaches the speed of light. It is this branch of physics that we will explore in the following pages.

Einstein did not name his idea "relativity," and the term is somewhat misleading. The word might cause us to think of moral relativity: the notion that ethical standards have no absolute authority. Or we might be reminded of relativism: the post-modern philosophy that there are no absolutes. Of course, these unbiblical views have *nothing whatsoever* to do with the branch of physics called "relativity." Actually, "invariance" probably better describes this field of physics. But the term "theory of relativity" has caught on and we will see that there are legitimate reasons why it was selected despite its ambiguity. So that is the term we will use in this text.

The theory of relativity is actually *two* theories; these are called *special relativity* and *general relativity*. They are not competing theories, but complementary ones. Special relativity deals with the nature of space, time, energy, mass, and so on in situations where gravity is negligible. General relativity is identical to special relativity except it includes the effect of gravity.[1] So we can think of special relativity as a subset or a special case of general relativity that is only for situations where gravity is negligible. This book will focus on special relativity, though we will touch on general relativity in later chapters.

The "Wow" Factor

There are two reasons why the physics of Einstein is absolutely, mind-blowingly awesome. First, relativity is *weird*, but in a wonderful way. The effects that "kick in" as an object approaches the speed of light are delightfully counterintuitive. They are very far removed from our everyday experience. For example, can you imagine being able to travel hundreds of years forward in time and meet your own distant descendants? From the physics of Einstein, we know that this is indeed possible under the right circumstances. Can you imagine a "star" that is so massive, that its own

---

[1] General relativity is far more difficult than special relativity, and uses a branch of mathematics/geometry that is seldom used outside the context of relativistic physics.

light cannot escape its gravity, rendering the star invisible? It's strange – but it's true!

I find it delightful that the Lord has constructed and upholds the universe in a way that challenges our preconceptions. It's as if the Lord has hidden "Easter eggs" throughout His universe for us to find and enjoy. Just when we think we have physics all figured out, we discover some new anomaly that forces us to reconsider our assumptions, and opens up a world of new possibilities. The physics of Einstein provides us with a wonderful glimpse into the mind of God.

The second reason involves the unusual way in which relativity was discovered and proved. Most branches of science involve a great deal of observation and experimentation. Even young students are familiar with the scientific method: we advance a hypothesis and construct an experiment that could potentially falsify the hypothesis. Observations from the result of the experiment either falsify the hypothesis or confirm it. If many such experiments are performed under a variety of different controlled conditions, and none falsify the hypothesis but do falsify the alternatives, then the hypothesis becomes a theory and we have some degree of confidence that it is *probably* a correct explanation or at least a good approximation of truth. Technically, we can never really know for certain that our theory is correct, because all it takes is one experiment to falsify it.

Most scientific knowledge is discovered in this way – but not relativity. *Einstein never did any physical experiment to test his theory.* And yet it is one of the most well-established theories of science. How is this possible? While most scientific discoveries are made by inductive, observational methods, relativity was discovered primarily through deductive logical reasoning. In the next chapter we will see how this works. My point here is that relativity, unlike most branches of science, is actually *provable*, within the context of its foundational premises.

Therefore, in the pages that follow, I will not only discuss the wonderful "weird" effects of relativity, but I will also prove that these effects *must* occur. The proofs will involve basic logic, basic geometry, and basic mathematics. By "basic" I mean anyone with a high school education should have no difficulty following the text. Of course, some people are intimidated by mathematics and will put down the book at this point. But I want to encourage such readers to proceed for the following two considerations.

First, logic, geometry, and mathematics are windows into the mind of God. Many people haven't been taught these subjects from a proper biblical perspective, and consequently have a negative impression of them. Yet, studying these topics allows us to experience the majesty of our Lord in a way that many people never get to experience. The proofs of relativity are elegant and beautiful. And most of them really aren't that difficult.

Second, a reader may choose to merely skim or skip the proofs and still enjoy reading about the strange effects of relativity. I have written this book with the proofs in boxed sections to facilitate such an approach. Some readers may skim the proofs to get the flavor, without going through the mathematical nuances. How you choose to read and enjoy this book is entirely up to you.

Since relativity is provable from its starting premises, we can have great confidence in this branch of physics. Consequently, the resulting technological advances that stem from relativity have been remarkably successful. The physics of Einstein has made possible the use of atomic/nuclear power. Relativity is necessary for global positioning systems (GPS) to work properly. It is used in particle accelerators. But some of the most interesting applications exceed our current level of technology. Beyond its practical applications, the physics of Einstein tells us something profound about the truly awesome mind of God.

Chapter 1

An Impossible Reconciliation

Sometimes the most amazing discoveries in science stem from an anomaly: something in the universe doesn't behave as expected. This leads scientists to reconsider their theories of how the universe behaves, and leads to new discoveries. The theory of relativity (the physics discovered by Einstein) is no exception. It was discovered largely as a result of an apparent contradiction between two different branches of physics. Let's briefly review these branches so that we can see the apparent conflict between them, and begin to understand how Einstein solved the problem.

First, we have the physics of Newton. Isaac Newton (1643 – 1727)[2] was one of the most brilliant men in recorded history. He developed the foundations for our understanding of the physics of motion and gravity. He made tremendous contributions to the field of optics (dealing with the reflection and refraction of light). He co-discovered the mathematics of calculus. Newton discovered the theoretical foundation for Kepler's laws of planetary motion. He designed and built the first practical reflecting telescope; many modern telescopes still use his basic design and are referred to as "Newtonians." He was also a Bible scholar and creationist.[3] Any one of his scientific achievements would be worthy of a Nobel Prize today. In summary, Newton was very, very smart.

---

[2] According to the modern Gregorian calendar

[3] Newton passionately studied the Bible and wrote a great deal on biblical topics. However, his views on the Trinity are the subject of some debate. They seem to have been unorthodox and possibly heretical.

The branch of physics concerning our topic and to which Newton heavily contributed is that of "classical mechanics" which deals with motion and energy. This branch of physics existed in its infancy before Newton. But his contributions to the field are so extensive that classical mechanics is often called "Newtonian physics." Newton was able to prove for the first time much of what others simply assumed, and he built upon these discoveries to develop what we now understand about the physics of motion. Newton's profound contributions to classical mechanics can be summarized by the three laws of motion and the law of gravity.

First among the laws of motion is the law of inertia: an object in motion tends to remain in motion unless acted upon by an outside force; and an object at rest tends to remain at rest unless acted upon by an outside force. The second part of this law is fairly intuitive. A rock doesn't just start moving on its own. It will only move if pushed or pulled by some external force.

The first part of the first law of motion may not be so obvious. In our everyday experience, moving objects tend to eventually come to rest. Such observations led the Greek philosopher Aristotle to conclude that moving things eventually come to rest unless a force is continually applied to them. After all, if you stop pushing on the rock, it eventually stops moving. Right? But Newton (and Galileo before him) realized that things moving along the surface of Earth only come to rest because the forces of friction and air resistance oppose their motion. In the absence of all forces, a moving rock will continue to move in a straight line at the same speed forever. It's an insightful conclusion, particularly when we consider that no human being has ever truly experienced an environment with no forces at all.

Newton's second law expresses the relationship between force, mass, and acceleration. Acceleration is a change in velocity. Mass is the amount of "stuff" and correlates with weight; e.g. massive things are very heavy (when on the surface of Earth). Force is when something is pushed or pulled. Newton's second law explains that the sum of all forces on a mass is equal to the product of that mass times the acceleration caused by the forces. The net force and the acceleration that follows from it are always in the same direction. Mathematically, this is expressed as $F = ma$. Pragmatically speaking, this means it takes a lot more force to move a heavy boulder than it does to move a pebble.

Newton's third law is perhaps the most quoted: "For every action there is an equal and opposite reaction." This law means that if something is pushed forward, then something else must also be pushed backward and with the same net force. Newton's third law is what makes walking possible. When you walk, your foot pushes backward on the Earth with some force. Consequently, the Earth pushes *forward* on you with equal force, which is why you start moving forward. Technically, when you walk forward, you cause the Earth to move backward just a bit. But since the Earth has so much more mass than you do, its acceleration backward is miniscule. (This follows logically from Newton's second law.) Newton's third law also allows rockets to work. The rocket pushes its exploded fuel downward, and the fuel pushes the rocket upward with equal force.

Finally, we have Newton's law of gravity. This law explains that all objects with mass produce a force of attraction (gravity) on all other massive objects. The more mass something has, the more gravity it has. Earth has a lot of mass, and the force of gravity near its surface is very apparent. This is what causes objects like apples to fall toward the Earth. Since you have mass, you have gravity too. But since you are much less massive than the Earth, your force of gravity on other objects is negligible. Nonetheless, if you were floating in deep space, and a rock was placed three feet away from you, it would slowly fall toward you due to gravity. (It would take nearly four hours to reach you.)

The force of gravity due to a mass diminishes with the square of the distance from that mass. So an object at some distance from the Earth will feel a pull of gravity from the Earth. If that object is moved so that it is now *twice* as far away from the Earth as it was, it will experience only *one-fourth* the gravitational pull it did previously. The force of gravity between two masses is expressed mathematically as $F = -GMm/r^2$. Here, the M and m are the respective masses of the two objects, r is the distance between them, and G is a constant, which is called the universal gravitational constant. The negative sign indicates that the gravitational force is attractive in nature.

Newton's laws of motion and his law of gravity make it possible to calculate how things will move through space: everything from rockets to asteroids. Newton was able to explain for the first time why planets orbit the sun the way they do. He realized that the Sun (since it has mass) must have gravity which deflects the paths of the planets from straight lines into ellipses. He computed the speed at which the planets must move to remain in their orbits. The success of classical mechanics is truly remarkable. These four simple laws can be written on a single index card. Yet, they are what made it possible for human beings to travel to the moon.

The Relative Nature of Motion

Implicit in all classical (Newtonian) physics is the fact that the motion of an object through space is only meaningful in comparison to some frame of reference. Motion is a change in position over the course of time. But a position is only meaningful when compared to a point of reference. Imagine calling someone on the phone and asking, "Where are you?" If the person answered "fifteen miles" this would make no sense. We would have to ask, "fifteen miles *from what?*" "Fifteen miles north of Denver" makes sense, and so does "fifteen miles east of Cambridge." But without a reference point, position is meaningless.

And motion, being a change in position over time, is therefore only meaningful when compared to something else. Imagine that the only thing in the universe is a single rock. We then ask the question, "Is that rock stationary or in motion?" The question is unanswerable because there is nothing with which to compare the rock. If the universe had *two* rocks, then we can ask if they are in motion relative to each other. However, without a reference frame, *motion* has no meaning.

The relative nature of motion can seem counterintuitive at first. This is because we have an enormous and obvious reference point beneath our feet that seems absolute: the Earth. When we measure speeds on Earth, it is almost always *relative to the surface of Earth*. When we see a speed limit sign "Speed Limit 50 MPH," few people stop to think, "50 miles per hour relative to what?" Obviously, the speed limit is relative to the surface of the Earth.

But the surface of the Earth is in motion relative to other things in the universe. The Earth itself rotates once every twenty-four hours. And so, when you are in bed asleep, you are actually moving at 700 miles per hour relative to the Earth's center.[4] Of course, so is your bed, and your room and your house and your neighborhood. These things all move together at the same speed, and so you don't perceive any motion. Relative to your bed, you are motionless; but relative to Earth's center, you are in motion.

---

[4] The exact speed depends on your latitude. Near the equator, the speed due to rotation is around 1000 mph, and goes to zero near the poles.

Why don't we feel it? If indeed we are on a rotating planet, moving eastward at 700 miles per hour, why do we not feel this velocity? The answer is that human beings cannot feel velocity. The only thing we can "feel" is a *change* in velocity. When the traffic light turns green and the car begins to accelerate, we feel a push backward into our seat. When we hit the brakes, we feel a push forward. When the car takes a sharp left turn, we feel a push to the right; when it takes a sharp right, we feel pushed to the left. But when the car is moving in a straight line at constant speed, we do not feel a push in any direction. We might feel bumps along the road that indicate our state of motion; but these are a temporary and sudden *change* in velocity. If you have ever flown on a large aircraft, when flying at constant speed through smooth air with no turbulence, you cannot tell if you are moving unless you look out a window. Human beings can only sense a change in velocity, not velocity itself.

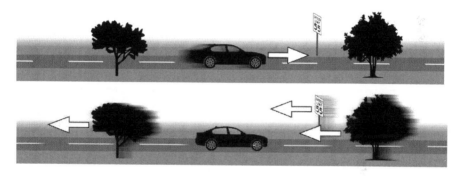

Furthermore, the velocity of anything is inherently relative to something else. The relative nature of motion means that you can always consider yourself to be stationary, with other things moving relative to you. When you drive along a stretch of highway at 55 miles per hour, it is fair to say that you are moving relative to the surface of Earth at 55 miles per hour (upper panel of the figure). But it is equally fair to say that you are stationary, and the Earth's surface is moving backward under your car at 55 miles per hour (lower panel). From the perspective of classical mechanics, both statements are equally legitimate and correct. This is a very important principle in the study of the physics of Einstein (and Newton as well), and so we must explore this a bit further and introduce some terminology.

Reference Frames

Since all motion is inherently relative, velocities are only meaningful when given relative to a frame of reference. A *reference frame* is the object (either a physical object or a conceptual one) to which the motion of another object is being compared. In our everyday experience, we almost always use the Earth's surface as our common reference frame, and so there is no need to mention it. The salesman who says, "This car can get up to 120 miles per hour," obviously means relative to the Earth's surface. After all, even a parked car is moving at 700 miles per hour relative to the Earth's center due to Earth's rotation.

However, in physics, we often imagine scenarios that take place in deep space, far from the Earth. In such situations it is always necessary to specify the reference frame if we are to discuss positions and velocities. We might imagine two people in two different spaceships floating past each other. In such a situation, it is convenient for each observer to take his own spaceship as the stationary reference frame by which all other motion is compared. Tom sees Pete's ship as moving north at 20 miles per hour (relative to Tom), whereas Pete sees Tom's ship as moving south at 20 miles per hour (relative to Pete). Each person is correct, according to his own reference frame. It would not make sense to ask, "Which person is *really* the stationary one – Tom or Pete?" Each is stationary relative to himself, and each sees the other person in motion. Both reference frames are equally legitimate.

There is an important caveat we must discuss before moving on. In classical mechanics, all reference frames are equally legitimate, if and only if they are *inertial* reference frames. An *inertial reference frame* is a non-rotating, non-accelerating frame. An inertial frame is therefore either stationary, or moving in a straight line at constant speed and not spinning relative to all other inertial frames. In our previous example, Pete is allowed to consider his own spaceship as a stationary reference frame *only* if the ship is not rotating, and not changing its speed or direction – and likewise with Tom. That is, each ship is simply drifting through space in constant straight-line motion. Under these circumstances, the laws of physics will work identically for both Tom and Pete. If they perform an identical scientific experiment in their respective spaceships – whatever that experiment is – they will get the same result. *The laws of nature work the same way in all inertial reference frames.*

Newton's laws of motion describe what happens with respect to any *inertial* reference frame. However, they do *not* describe what happens with respect to non-inertial frames. Consider the following example. From the first law, we know that an object at rest remains at rest if there are no outside forces. Imagine driving along a straight stretch of highway at constant speed. This is an inertial frame. Then suppose we place a marble on a flat, empty seat in the car. The marble remains stationary relative to the car, as we would expect from Newton's first law. However, when we step on the brakes, the car decelerates – it is no longer inertial. Will the marble remain stationary relative to the car? No. It will roll forward even though no external force acted upon it. This would seem to violate Newton's first law until we recognize that the car is now a non-inertial reference frame in which Newton's laws do not apply as written.[5]

Newton's laws *do* allow for rotation and acceleration of objects and can correctly describe the motion due to forces. However, we must not treat rotating or accelerating objects *as inertial reference frames* from which to measure the motion of other things. So when we use Newton's laws to do a calculation, we must measure all velocities with respect to an inertial reference frame. It does not matter which frame we use – as long as it is an inertial one we will get the correct result.

Non-inertial reference frames can be very convenient and useful, and there is nothing wrong with using them. We just need to remember that Newton's laws do not correctly describe motion with respect to a non-inertial reference frame. So if we record the path of some object with respect to a non-inertial reference frame, we must convert our observations to an inertial frame before applying Newton's laws. We can then convert back to a non-inertial frame if desired.

The Earth is a non-inertial reference frame. This is because it is rotating. And even if it weren't rotating, it is traveling around the sun in an elliptical path – not straight-line motion. Yet, it is very convenient to use the Earth as a reference frame since we all live on its surface. But if we wanted to use Newton's laws to send a rocket from the Earth to the moon, we would have to do our calculations in an inertial frame. Of course, the Earth rotates and revolves slowly enough that it can be approximated as inertial in some situations. For the sake of clarity, in this book we will pretend (unless otherwise noted) that the Earth is not rotating or orbiting

---

[5] There is a way to use Newton's laws in some non-inertial reference frames. By introducing two fictional forces, Newton's laws will continue to work as written with respect to a rotating frame. These forces are "centrifugal force" (not to be confused with *centripetal force*, which is real) and the "Coriolis force." It is often convenient to introduce these fictional forces which allow Newton's laws to correctly describe motion with respect to a rotating frame, such as the Earth.

the sun, but is (nearly) inertial, as if it were floating through space at a constant speed and direction, and not rotating.

In summary, the physics of Newton (classical mechanics) is a remarkably successful branch of science. It explains why objects in the universe move as they do. From golf balls and airplanes to planets and moons, the three laws of motion and law of gravity can explain it all. And most importantly, we learn that all inertial reference frames are equally legitimate. There is no such thing as an absolute velocity, because all motion is relative. The observed velocity of any object will depend on the reference frame of the person making the observation. The person in the passenger seat is stationary relative to the driver of the car, but is moving at 55 miles per hour relative to someone sitting on the side of the road. This seems straightforward and obvious. Yet, it seems to clash with what comes next.

Maxwell's Equations

In the mid 1800s a Scottish physicist named James Clerk Maxwell made a discovery that put a chink in the armor of Newtonian physics. A devout Christian and biblical creationist, Maxwell is considered one of the greatest physicists of all time. His area of interest dealt with electric and magnetic forces. Physicists already knew that electricity could be used to produce a magnet, and magnets in motion can produce electricity. But the exact relationship between these forces was not understood. Maxwell figured it out. His research in this area culminated in four equations that correctly describe the nature of all electric and magnetic phenomena.

Like Newton's laws of motion, Maxwell's equations are simple enough to be written on a single index card, and yet powerful enough to explain all kinds of phenomena. However, unlike Newton's laws, Maxwell's equations are not easy to explain without a strong background in calculus. For this reason, I will discuss only the results of these equations that are pertinent to our discussion of relativity.

Surrounding every charged particle is an invisible electric field. You can think of an electric field as an invisible "fog" that surrounds all charged particles, and which gets progressively thinner with increasing distance. This field exerts a force on any other charged particle placed within it. Likewise, surrounding every magnet is a magnetic field. It is also normally invisible, but there are ways to see it. For example, if you have ever seen iron fillings sprinkled around a magnet, the arcs they form map the structure of the magnetic field. Magnetic fields and electric fields are different, but they are mathematically connected.

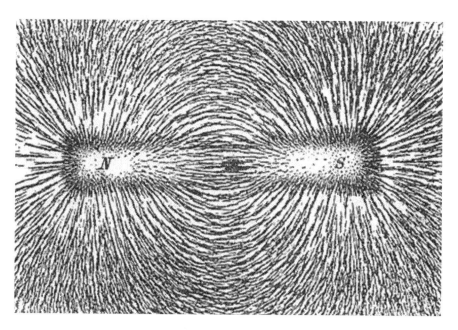

Maxwell's equations show that when a magnetic field changes in any way, it will temporarily create an *electric* field surrounding the location of the change. You can change the magnetic field in space by, for example, moving a magnet. Likewise, any change in an electric field will temporarily create a magnetic field in the surrounding space. You can change an electric field by moving a charged particle, like an electron. Maxwell's equations describe how electricity (moving electrons) produces magnetism, and how moving magnets across a wire can induce electricity. The applications are endless.

Maxwell's equations also show something particularly interesting. Since a charged particle has an electric field surrounding it, we can create a changing electric field by wiggling that particle up and down. This creates a magnetic field. But since we are continually changing the direction of motion of the electron, the magnetic field will also be constantly changing from positive to negative – and so it will create another electric field. But this electric field will also oscillate for the same reason, giving rise to another magnetic field and so on forever. This chain reaction propagates through space as follows:

A-B-C-D-E-F-G-H

We start moving the electron up and down at point A, which creates a changing magnetic field at B, which induces a changing electric field at C, which induces a changing magnetic field at D, which induces a changing electric field at E, and so on. The oscillating electric and magnetic field propagates away from its point of origin, like an undulating wave out into space – an electromagnetic wave.

Thus, by rapidly changing the direction of electricity in a wire, we can create an electromagnetic wave that travels away from the wire through the vacuum of space. If the wave reaches another wire in space, it will induce a changing electric field in that second wire that matches the frequency of oscillations in the first wire. This is how radio works. Sound is converted to electric oscillations in a wire (that is the transmitter), which radiates electromagnetic waves into space. When these waves intercept the wire on the receiver, they induce electric oscillations that we can convert back into the original sound. Radio waves are electromagnetic waves.

Maxwell's equations also show that these electromagnetic waves must travel through vacuum at a very specific speed. This speed is related to two constants in nature which set the strength of the relationship between electric fields and magnetic fields and are called the permittivity of free space (ε_0), and the permeability of free space (μ_0) respectively. Maxwell found that the speed of all electromagnetic waves is equal to $1/\sqrt{\varepsilon_0 \mu_0}$. This turns out to be very fast, and is roughly equal to 186,000 miles per second! Amazingly, this happens to be the speed of light.

Although physicists of the 1800s did not know the speed of light as precisely as we do today, they did know its approximate speed. Maxwell realized that the estimated speed of light was essentially the same as what his equations showed must be the speed of all electromagnetic waves.

Maxwell drew the reasonable inference: *light is an electromagnetic wave.* Light is the same type of thing as radio – it is an electromagnetic traveling wave caused by an oscillating electric field. Visible light simply has a much higher frequency than radio. But they are the same "substance."

It was a truly remarkable discovery: Maxwell reasoned that light is an electromagnetic wave that must travel through vacuum at approximately 186,000 miles per second, a speed that physicists denote by the lowercase letter 'c.' Today, we know this value even more precisely: c = 186,282.3974 mi/s. Maxwell's equations have been remarkably successful in predicting the outcome of experiments involving electricity and magnetism. They correctly describe the characteristics of light and radio. Their usefulness rivals that of Newton's laws of motion. Yet, a problem remains. This remarkable result of Maxwell's equations flatly contradicts an important aspect of Newton's laws of motion.

The Paradox

We saw previously that Newton's laws of motion describe how force, mass, acceleration, and velocities behave for any object as measured by any inertial reference frame. We saw that motion is inherently relative, that speeds are only meaningful when the reference frame is specified. Two different observers that are in motion relative to each other will record different velocities for the same object. The driver of a car sees his passenger as stationary relative to himself; but the person on the sidewalk sees the passenger moving (along with the car) at 55 miles per hour. The passenger is stationary in one reference frame, and moving in a different reference frame. Speeds are not absolute, but depend on the reference frame of the observer. This is intuitively obvious. But according to Maxwell, this cannot be true.

Maxwell's equations indicate that there is an absolute speed: the speed of light. Recall that Maxwell's equations imply that all electromagnetic waves must travel through vacuum at speed c. No reference frame was specified; Maxwell's equations imply that this must be true for *any* (inertial) reference frame.

So on the one hand, we have Newton's laws of motion which imply that all speeds are relative and reference-frame dependent. And on the other hand, we have Maxwell's equations which imply that light and all other electromagnetic waves must travel at exactly c (186,282.3974 mi/s) *regardless of the reference frame.* We have two very successful branches of science, each of which makes correct predictions in their respective areas. And yet they are contradictory.

Newton:	There is no absolute speed/velocity; the measured speed of anything depends on the observer's reference frame.
Maxwell:	The speed of light is absolute; the measured speed of light does _not_ depend on the observer's reference frame.

To emphasize the strangeness of this paradox, consider the following scenario. Imagine a world-class baseball pitcher who is able throw a ball at a staggering speed of 100 miles per hour. While riding in a car at 50 miles per hour, he rolls down the window and throws the ball forward at 100 miles per hour faster than the car. A person standing on the sidewalk would see the ball moving at 150 miles per hour. That's what Newtonian physics predicts, and everyday experience confirms it.

But this doesn't work with light. When the driver moving at 50 mph turns on his headlights, he sees the light moving at speed c (186,282.3974 miles per second) faster than himself. We would think that the observer on the sidewalk would see the light moving at a slightly faster speed: $c + 50$ mi/hour. But she doesn't. She sees the same light beam moving at exactly speed c relative to the sidewalk.

The absurdity becomes even more apparent when we increase the velocities. Suppose the car were traveling down the road at half the speed of light: $\frac{1}{2}c$. The driver turns on his headlights, and measures the light as being speed c relative to his car. We would certainly be inclined to think that an observer on the sidewalk would measure the light at $1\frac{1}{2}c$. But she doesn't. She finds the light to be moving at speed c relative to the sidewalk, which is only $\frac{1}{2}c$ faster than the car.

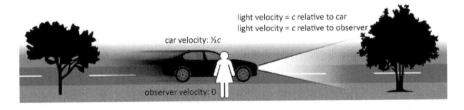

light velocity = c relative to car
light velocity = c relative to observer
car velocity: $\frac{1}{2}c$
observer velocity: 0

How can this possibly be true? How can two different observers both measure the same beam of light as speed c faster than themselves when they are moving relative to each other? This paradox perplexed 19th century physicists. Albert Einstein solved the paradox. In a flash of creative insight, he realized that physicists had made a critical assumption – one that seems perfectly reasonable on the surface, and yet one that happens to be false.

Chapter 2

Resolving the Paradox

Intuitively, we expect that velocities should add. That is, the speed of a car relative to the road plus the speed of the baseball thrown from the car should equal the speed of the baseball relative to the road. Physicists call this principle the *Galilean velocity addition theorem*. It is represented as follows:

$$W = v + w \qquad\qquad (2.1)$$

Here, *v* is the velocity of the car relative to the road, *w* is the velocity of the baseball relative to the car, and *W* is the velocity of the baseball relative to the road. We applied this formula in the last chapter when the car moved at 50 miles per hour, a baseball was thrown from it at 100 mph, and the speed of the ball relative to the road was 150 mph.

The Galilean velocity addition theorem works perfectly well for cars and baseballs. But it fails miserably when we try it with a beam of light. Light always moves at speed *c* regardless of the reference frame – not *c+v*. Maxwell's equations imply an absolute speed of light for any inertial observer. But how can this be?

The success of Newtonian physics was well known in the 19th century, and few doubted that it needed any modification. Most physicists believed that Maxwell's equations, though remarkably successful in their predictions of electromagnetic phenomena, would need to be modified to align with Newtonian physics. Or, Maxwell's equations were thought to be true *only* for a particular reference frame – the reference frame of the *medium* of light.

A 'medium' is the substance through which a wave travels. Sound, for example, is a compressional wave traveling through air. Since sound is the oscillation of the molecules of air, sound cannot exist without air. In fact, all waves that were well-studied by the 19th century required some sort of physical medium in which to travel. So naturally, many physicists assumed that electromagnetic waves like light and radio required some material substance in which to travel. They referred to this hypothetical substance as the *luminiferous ether*. This, they assumed, must be the reference frame in which light travels at speed *c*.

Since we are able to see distant stars, physicists inferred that the luminiferous ether must fill all of space. The Earth must travel through this ether in its annual path around the sun. And since we do not feel any 67,000-mph wind, physicists further assumed that the ether had the ghostly ability to pass harmlessly through ordinary matter. Nonetheless, as the Earth rotates on its axis, and as it orbits the sun, our direction of motion relative to the ether should gradually change. And if light always travels at speed *c* relative to this ether, this means we should detect a very small change in the speed of light relative to the Earth. This follows from the Galilean velocity addition theorem.

In 1887, physicists Albert Michelson and Edward Morley performed an experiment to test this hypothesis. Their experiment was ingenious. They used the "interference" property of the wave nature of light to measure any infinitesimal change in the speed of light relative to Earth's surface. Such tiny changes were expected to occur as the Earth changes its direction in its daily rotation and its annual orbit relative to the luminiferous ether.

But the experiment failed. There was absolutely no change in the speed of light relative to Earth's surface in Earth's daily rotation or in its annual orbit. Contrary to the expectations of Newtonian physics, the speed of light remained *c* relative to the surface of Earth and *not* relative to some undetectable ether. Maxwell's equations were right. The speed of light is always exactly *c* in a vacuum *relative to any observer*, regardless of the motion of that observer.

I cannot overstate the strangeness of this phenomenon. No matter how fast you move, light is always exactly *c* faster than you. It is almost as if the light "knows" what the observer is doing, and then adjusts accordingly. If you move away from the Earth at 99% of the speed of light, you will find that light is still exactly *c* faster than you. Yet, an observer on Earth will see the same beam of light moving at *c* relative to Earth. How can this be?

The Solution

To understand the solution to this paradox, let us phrase the problem in a slightly different way. Suppose Michael invents a new type of car that can travel down a road at 50% the speed of light. Sarah watches from a stationary position along the side of the road. When Michael speeds down the road and turns on his headlights, he measures the light as travelling at speed *c* faster than his car.

Since Michael travels at half the speed of light, his speed is about 93,141 miles per second. He measures the light from his headlights as 186,282 miles per second faster than his car. Sarah sees the light as moving 186,282 miles per second relative to her stationary frame. Therefore, from her perspective, the light is moving only 93,141 mi/s *faster than Michael*, since he is already moving at 93,141 mi/s. So how can Sarah measure the light beam as 93,141 miles per second faster than Michael, whereas Michael measures the light as moving 186,282 miles per second faster than himself? They are both measuring the speed in the same units – miles per second. Yet, they get a different answer.

Einstein realized that the only reasonable inference was to conclude that Michael's miles or seconds are not the same as Sarah's miles or seconds. The speed of something is the distance that it covers divided by the time it takes to cover that distance. So to correctly measure a speed, we must be able to measure distance and time accurately. Logically, if Michael and Sarah arrive at a different speed for the same beam of light (in terms of how much faster it is than the car, either 93,141 mi/s or 186,282 mi/s), it follows that they must be measuring either distance or time (or both) differently.

Let me illustrate with an example. Suppose Samantha claims to be able to run the one-hundred-yard dash in only six seconds. That would make her speed about 34 miles per hour – impossibly fast for a human being. Her friend Jake is naturally skeptical and challenges her to prove it. They mark off a path one hundred yards in length. Samantha and Jake each have a stopwatch. "On your mark, get set, go!" Both start their stopwatch as Samantha bursts forth in an all-out sprint. Both stop their watch when she reaches the finish line. Sure enough, Samantha's stopwatch records six seconds. But Jake's stopwatch records twelve seconds. If Jake's time is correct, then this means that Samantha's true speed is only 17 miles per hour – a much more realistic value. Samantha had computed an incorrect speed because *her stopwatch is running slow*. It recorded only six seconds, when in fact twelve seconds had elapsed.   This caused her speed calculation to be wrong.

There is another way in which Samantha might have been mistaken, even if her stopwatch had been working correctly.   The tape measure she used to mark off one hundred yards may have been inaccurate – too short.   It says the distance is one hundred yards, when in fact the distance is only fifty yards.   This will make Samantha's speed seem twice as fast as it really is.   Samantha only thinks she can run one hundred yards in six seconds, when she is actually running only fifty yards in six seconds.

Einstein recognized that something like this must be happening in order for all (inertial) observers to see the speed of light as always exactly c faster than themselves.   If Michael, speeding down the road at 93,141 miles per second measures the speed of light as 186,282 mi/s faster than his car, and yet Sarah measures the same light beam as only 93,141 mi/s faster than Michael's car, then obviously Michael and Sarah have either a different standard for "miles," for "seconds," or for both.

To be clear, it is *not* the case that Michael's stopwatch or tape measure is inaccurate *when stationary*.   Indeed, when he is stationary, his stopwatch ticks at the same rate as Sarah's.   And his tape measure lines up perfectly well with Sarah's.   Yet, when Michael is speeding down the road, his measurement of speed differs from Sarah's.   In other words, it is *only when Michael is moving* that his clocks and/or measuring tapes differ from Sarah's.   Einstein recognized that *motion itself* must affect either the rate at which clocks tick, or the length of measuring tapes – or possibly *both*.   This is the only way in which all observers can see the same beam of light as moving at the same speed relative to themselves.   Either moving clocks tick slower than stationary ones, or moving rulers are shorter than stationary ones.   It is hard to believe that simple motion could affect that rate at which a clock ticks or the length of a ruler.   Yet, at least one of these *must* occur.

Let's consider some numbers to see how either one of these options would resolve the paradox.   Suppose Michael's clock slows down such that when he is moving at half the speed of light his clock ticks only one second when Sarah's clock ticks two seconds.   From Sarah's point of view in two seconds, the light beam has traveled 372,564 miles – but Michael has traveled 186,282 miles in that same time.   So the light has gotten ahead of his position by 186,282 miles in two seconds.   Sarah computes that the light is 93,141 miles per second faster than Michael.   But in this time, Michael's clock has ticked only one second.   So from his point of view, the light has traveled 186,282 miles in only *one* second, and its speed is therefore 186,282 miles per second faster than himself. This would resolve the paradox.

Alternatively, if Michael's clock is unaffected by motion, then perhaps his measure of distance is affected. Suppose that when he is moving at half the speed of light, Michael's tape measure shrinks to exactly half its original size. In one second, Sarah measures the light as moving 186,282 miles, whereas Michael has moved 93,141 miles in that same time. The light is therefore ahead of Michael by 93,141 miles in that second. Sarah computes that the light is moving faster than Michael by 93,141 miles per second. But Michael's tape measure has shrunk to half its original length. So in that same one second, he measures the light to have gotten ahead of him, not by 93,141 miles (as measured by Sarah's tape measure), but by twice that amount (186,282 miles). So he again concludes that the light is 186,282 miles per second faster than himself. This again resolves the paradox.

So at high speed, it must be the case that either clocks slow down or rulers shrink. Either option seems bizarre, and contrary to our everyday experience. So we are naturally reticent to accept such a conclusion. Obviously, motion does not seem to affect the passage of time or the length of any object. A moving car doesn't seem to be any shorter than a stationary one. Right? And the clock in a moving car seems to tick at the same rate as a clock of identical construction in a stationary car. But since light is very fast, only a very slight difference in clock rates or lengths between moving and stationary frames would be necessary to keep the speed of light constant for all inertial observers. So we might not notice these effects in our everyday experience.

But the Michaelson-Morley experiment confirmed that at least one of these two effects *must* occur in reality, regardless of our intuitive expectations. Motion must affect either the passage of time or the length of objects or both. Thus, lengths and/or times are not "absolute" in the sense of being the same for all inertial observers. Instead, one or both are *relative* to the velocity of the observer. The theory of relativity gets its name from this fact.

Revealing the False Assumption

Before Einstein, virtually everyone assumed that the passage of time and the length of objects are absolute and not affected by velocity. But Einstein realized that at least one of these assumptions is false. Counterintuitive though it may seem, either sizes or times are affected by motion.

This is why the Galilean velocity addition theorem doesn't work when the two velocities being added are measured in two different reference frames. The velocity v is measured by someone with a different measure of time or length than the person measuring the velocity w. Therefore, we cannot add the two velocities until we convert one of them into the same system as the other. By analogy, we would not add one inch to one centimeter and expect to get two inches. Instead, we must first convert one of the length measurements so that both use the same standard of units, e.g. 1 centimeter = 0.3937 inches. So the resulting length is 1.3937 inches, not two inches.

To be clear, the Galilean velocity addition theorem does work when the two velocities are measured by the *same* observer. Then the same clocks and measuring tapes are used to measure all velocities, so no conversion is necessary. Therefore, if I observe a car traveling at ½c, and I also observe light from the headlights moving at ½c faster than the car, the total speed of that light beam will indeed be c. But when the two velocities are made by *different* observers, the speeds will not simply add since those two observers measure either lengths or times differently. In that situation we will need to use the relativistic velocity addition theorem, which we will derive in a later chapter.

What remains is to discover which of these two effects is responsible for keeping the speed of light the same for all inertial observers. Is it that moving clocks tick at a different rate from stationary clocks of identical construction? Or is it that moving rulers are different lengths from stationary ones of identical construction? Or do both effects occur? Albert Einstein discovered the answer. But what's even more amazing is the *way* in which he discovered it.

Chapter 3

The Thought Experiment

We've seen that the speed of light in vacuum is constant in all inertial reference frames as anticipated by Maxwell's equations. Albert Einstein recognized that this would only be possible if the rate at which clocks tick and/or the lengths of rulers are not absolute but change with velocity. But which effect is responsible for the constancy of light – the shrinking of rulers or the slowing of clocks? You might think that Einstein at this point would have constructed a physical experiment to measure lengths of moving rulers, or rates of moving clocks to discover the answer. But he didn't. Einstein was able to figure out the answer without conducting any physical experiment whatsoever. Instead, he used a method called a *Gedankenexperiment*.

A Gedankenexperiment, or "thought experiment," is a procedure in which a conceptual experiment is performed only in the mind, but not physically. The results of the experiment can be derived through pure logical reasoning. We need not physically perform the experiment; yet we know what would happen due to logical principles. Gedankenexperiments are very similar to mathematical, geometric, or logical proofs.

Einstein was not the first person to make use of such a method. Galileo used a Gedankenexperiment to prove that all objects will fall at the same rate (in the absence of air) regardless of their mass. This runs contrary to Aristotle's claim that heavier objects fall faster than lighter ones. Galileo's proof took the form of a *reductio ad absurdum*; this is an argument in which we temporarily assume the *opposite* of what we wish to prove and show that it leads to a contradiction. Since contradictions cannot be true, the initial assumption that led to the contradiction is falsified. Here is the essence of Galileo's proof:

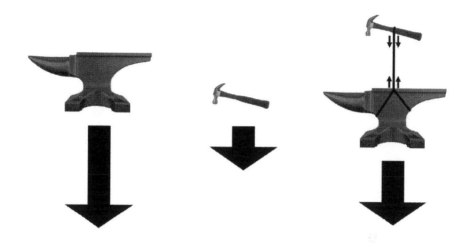

Assume for the sake of argument that heavy objects fall faster than light ones. In this case, an anvil will fall faster than a hammer, since the anvil is the heavier object, as illustrated in the figure. Now tie the anvil to the hammer with a short rope and drop them both. Since the anvil falls faster than the hammer, it will be at the bottom of the rope with the hammer at the top. The anvil will pull down on the hammer and the hammer will pull up on the anvil, slightly slowing its fall. Thus, the combination of the anvil and hammer will fall *slightly slower* than the anvil alone. But the anvil and hammer combination is necessarily *heavier* than the anvil alone, so it should fall faster according to our starting assumption– not slower.

The assumption that heavier objects fall faster than lighter ones leads to the contrary conclusion that heavier objects fall *slower* than lighter ones. Therefore, that initial assumption must be false. The outcome would also be contradictory if we assumed that lighter objects fall faster than heavier ones. Since both of these assumptions lead to a contradiction, they cannot be true. The only remaining possibility is that all things fall at the same rate in the absence of air, regardless of their mass.

In this example, there was no need for Galileo to actually drop anything, or measure the rate at which anything falls. He reached the correct conclusion through pure reason.[6] And it's quite conclusive. Of course, physical experiments will confirm the result.[7] And they have. Apollo 15 commander David Scott dropped a hammer and feather on the moon to show that indeed they fall at the same rate in the absence of atmosphere.

The Two Postulates of the Theory of Relativity

In a similar fashion, Einstein was able to derive the theory of relativity without performing a single experiment. He required only the basic definitions of various physical properties, such as mass and velocity, and he required two starting postulates. A postulate is something that is assumed to be true without absolute proof, is usually very intuitive and consistent with all observations, and forms the basis from which other truths may be derived. The two postulates of the physics of Einstein are the following:

1. The relativity principle: the laws of physics are the same in all inertial reference frames. This principle implies that if you do any experiment under a given set of conditions, the outcome will not depend on your velocity. For example, if you are stationary relative to the surface of Earth, you find that water boils at 100° Celsius at sea level. If you do the same experiment on a fast boat under otherwise identical conditions, you'll get exactly the same result. The relativity principle goes back to the relative nature of motion we discussed in an earlier chapter. There is no absolute velocity frame. Things are in motion only when compared to something else. The relativity principle informs us that no velocity frame is preferred over any other with respect to how the laws of physics work. Note that this applies to inertial frames only – non-accelerated, non-rotating frames.

2. The invariance of the speed of light: the speed of light in vacuum is always exactly c (186,282.3974 miles per second) with respect

---

[6] This can also be proved mathematically from Newton's second law and the law of gravity. Substitute the second law as the force into the formula for gravity, and divide both sides by m. The resulting acceleration due to gravity is not a function of m (the mass of the falling object). Thus, all things fall at the same rate regardless of mass.

[7] There is a claim that Galileo actually performed such an experiment by dropping two objects of different mass from the leaning Tower of Pisa. However, this claim is disputed.

to any inertial reference frame. It doesn't matter how fast you are moving. You will always measure the speed of light as c faster than yourself.

These are both postulates in the sense that they cannot be definitively proved to be true in all circumstances. This is because we do not have universal experience (no person has performed every possible experiment at all locations and times in the universe). But make no mistake: we do have good reasons to believe that these postulates are true. All experiments that have been performed to date have shown that the speed of light in vacuum is indeed always exactly c relative to any inertial observer. Likewise, all experiments performed so far show identical outcome regardless of velocity − confirming the relativity principle. The two postulates of relativity are as well-confirmed as any scientific inference can be. Once we accept these postulates, the rest of relativity can be logically deduced from them quite conclusively through the use of the Gedankenexperiment.

The Fast-Moving Train

For the rest of this chapter and the next several chapters, we will consider a thought experiment that is essentially the same one that Einstein used to discover the theory of relativity. We imagine two different observers that are in motion relative to each other. Michael is riding on a high-speed train. Sarah is watching, stationary relative to the surface of Earth. We imagine a coordinate system attached to Sarah in which she is always at the origin. And let's define the motion of the train to be in the positive x-direction at some high velocity v. The train does not move in height or take any turns, so the motion along the y-axis and z-axis is zero.

Michael also has a coordinate system with himself at the origin at all times. From his point of view, the train is stationary, and the Earth and Sarah along with it are moving backward relative to him at high speed in the negative x-direction at velocity $-v$. From the relativity principle, we know that Michael experiences physics exactly the same way as Sarah. And both see the speed of light as being exactly c relative to themselves. The coordinates that Sarah assigns to a given object are x, y, z and time t. For Michael, we'll use the coordinates $x', y', z',$ and t'. The superscript symbol is a "prime" mark, and so we verbally refer to Michael's coordinates as "x prime, y prime, z prime, and t prime."

Both Sarah and Michael agree to set their clocks to $t = t' = 0$ at the moment Michael passes by Sarah. We will ignore the small distance Sarah is standing from the train track. So at time $t = 0$, Sarah and Michael are in approximately the same place: $x = 0$, $y = 0$, $z = 0$, $t = 0$ by Sarah's coordinate system, and $x' = 0$, $y' = 0$, $z' = 0$, $t' = 0$ by Michael's coordinate system.

Since the train moves only in the x-direction, from Newtonian physics we would expect Michael's non-x coordinates would be exactly the same as Sarah's. That is $y' = y$, $z' = z$, and $t' = t$. And we would expect that $x' = x\text{-}vt$. These four simple equations are called the Galilean transformation equations. And they work pretty well as a good approximation for things traveling much slower than light. But we know they cannot be exactly right, because they assume that length and time are not affected by motion.

From the previous discussion we know that either Michael's clocks tick slower than Sarah's or Michael's rulers are shorter than Sarah's – or possibly both. And so, we may not assume that $t' = t$ or that $y' = y$ and so on. We saw in the previous chapter that if moving rulers shrink in the x-direction (the direction of motion), this would explain the constancy of the speed of light. But if rulers can shrink in the direction of motion, perhaps they can shrink in other directions as well. So we must now ask the question: do rulers moving in the x-direction shrink or expand in the y or z direction, or are they unchanged?

We will answer this question with a Gedankenexperiment using the relativity principle. It will be a *reductio ad absurdum*, similar to Galileo's thought experiment above. That is, we will start with an assumption and find that it leads to a contradiction – proving that the assumption must be false. I got this example from another physicist and thought it illustrated the point really well. I call it the "painting the hallway" experiment.

Painting the Hallway

As Michael speeds down the tracks in his high-speed train he comes to a narrow pass – a hallway. Sarah has measured the hallway to be fifteen feet wide from her stationary reference frame. The train, when stationary, is

only ten feet wide. So there are two-and-a-half-feet clearance on either side. The hallway is a dull grey color, and Michael thinks it would look much better with red strips. So he decides to paint the hallway as his train passes through. He attached two brushes, dripping with red paint on opposite ends of a long pole so that they measure exactly 15 feet from the tip of one paintbrush to the other. Michael places them such that they are parallel to the y-axis; they are sticking out the open windows across the width of the train, each paintbrush extending two-and-a-half feet beyond the width of the train. As the train passes through the narrow hallway, the brushes will exactly touch the sides of the wall, as shown in the figure. They will paint a nice red stripe on each side of the hallway as the train passes by. Or will they?

We already know that, because he is in motion, Michael's rulers are either shorter than Sarah's or his clocks tick slower, or both. Let's assume that Michael's rulers are shorter than Sarah's in the y-direction (width) when his train is moving in the x-direction. In other words, we assume that widths contract at high speed. If this is the case, then the two paintbrushes that are fifteen feet from tip to tip as measured by Michael, will be less than fifteen feet as measured by Sarah since Michael's rulers are shorter in the y-direction than Sarah's. But the hallway is fifteen feet wide as measured by Sarah. Therefore, Michael's stick will not be wide enough for the paintbrushes to paint the hallway; they will be less than fifteen feet tip to tip as measured by Sarah. They will miss, and the hallway will not be painted, as shown in the second figure.

From the relativity principle, Michael is allowed to consider himself and the train as stationary, with the rest of the world moving by at high speed. From his perspective, it is the *hallway* that is moving in the negative *x*-direction. From his perspective, Sarah is in motion. And if our starting assumption is correct, that moving rulers contract in the *y*-direction, then Sarah's rulers are contracted relative to Michael's. If so, then when Sarah measures the hallway to be fifteen feet across, the true width by Michael's rulers will be *less* than fifteen feet across. In that case, his paintbrush stick will be more than wide enough to paint the hallway, since it is slightly less than fifteen feet across from his reference frame, as illustrated in the third figure.

From Sarah's point of view, the hallway does not get painted. But from Michael's point of view, the hallway *does* get painted. This is a contradiction because both Sarah and Michael can go back to the hallway and objectively check. Either it will be painted or it won't be, but not both. If widths are affected by motion, then this leads to a contradiction. *Therefore, we must conclude that widths are not affected by motion.* That is, motion along the *x*-axis does not affect the width of a ruler along the *y*-axis.

And by a similar thought experiment in which we paint the ceiling and floor of a tunnel, we find that the *z*-axis is also unaffected by motion along the *x*-axis. By this thought experiment, we know that moving trains have the same height and width as their stationary counterpart. This further means that the y' and z' coordinates that Michael measures for any event will match the y and z coordinates that Sarah measures for the same event: $y = y'$ and $z = z'$.

So our first thought experiment in relativity has a rather boring outcome. There are no strange or wonderfully counterintuitive effects that happen in the *y* or *z* dimension when an object moves along the *x*-axis. Widths and heights neither shrink nor expand with motion. As boring as this result is, we will need this result in order to prove other things about relativity.

But what about lengths along the x-axis and the passage of time? We know that at least one of these must be affected by motion in order for the speed of light to be the same in all inertial frames. This is where relativity starts to get really interesting.

Chapter 4

Time Dilation

We saw in the previous chapter that heights and widths are not affected by motion along the x-axis. And so Michael (a passenger on board a high speed train) and Sarah (a bystander watching the train) will agree on the height and width of the train. We found that this must be the case by conducting a Gedankenexperiment – a "thought experiment" in which we analyzed what must logically happen from Michael's reference frame and Sarah's frame respectively.

What about the passage of time? Does Michael's moving clock tick at the same rate as Sarah's stationary clock? We again use a thought experiment, along with the two postulates of relativity. Recall that these are (1) the laws of physics are the same in all inertial reference frames, and (2) the speed of light is always c (in vacuum) with respect to any inertial observer.

In this thought experiment, Michael is again on a high-speed train, while Sarah is stationary and observes from the side of the track. Michael has brought with him a flashlight, a mirror and a stopwatch. He attaches the flashlight to the floor of the train, with it aimed toward the ceiling above. And on the ceiling, directly above the flashlight, he attaches the mirror so that it will reflect the light directly back to the flashlight. Michael then lies down on the floor next to the flashlight, facing upwards so that he can see the reflection of the flashlight in the mirror directly overhead.

The goal of this experiment is to measure how long it takes for the light to leave the flashlight, bounce off the mirror and return to the floor. So Michael will turn on the flashlight at the same time that he starts his stopwatch. He will hit "stop" as soon as he sees the reflection of the light in the mirror. In reality, this would take place so quickly that Michael could never measure the time accurately by eye. But let's imagine that the train is tall enough, or that light is slow enough that Michael can accurately use his stopwatch to compute the amount of time it took for the light to go from the floor, reflect off the ceiling, and return to the floor.

This experiment takes place on a train that is moving at high speed relative to Sarah. She watches the experiment from her stationary position. She uses her stopwatch to measure the amount of time it takes for the beam to leave the flashlight, reflect off the mirror and return to the floor. If the flow of time is unaffected by motion, then Sarah and Michael will record the same time for this experiment. However, if Michael records a different time than Sarah does, then this will prove that his clock ticks at a different rate from hers. *It will prove that motion affects the passage of time.*

Using the postulates of relativity, we can compute how long it takes light to leave the flashlight, bounce off the mirror, and return to the floor, first from Michael's perspective and then from Sarah's. We do not need to actually physically do the experiment. From geometry, we can know what the results would be. Here is the proof:

We will define our variables as follows: v = the velocity of the train relative to the track, h = the height of the train, T' = the time Michael records for the light to return to its origin, T = the time Sarah records. From Michael's point of view (left side of the figure), he and the train are stationary; the rest of the world is moving past him at high speed. And from the second postulate of relativity, the speed of light is exactly c from his perspective. This light must travel a total distance of $2h$ since it must travel to the ceiling and back. And since the distance something travels ($2h$) is equal to the velocity times the time (cT'), this means $2h = cT'$. Solving for T' we find that Michael records the time for the light to return as:

$$T' = 2h/c$$

From Sarah's perspective (right side of the figure), the experiment looks a bit different because the train is moving. But one thing remains the same: h – the height of the train. Recall from the "painting the hallway" thought experiment from the previous chapter, motion in the x-direction does not affect width in the y-direction or height in the z-direction. Thus, Sarah and Michael both measure the same height for the train: h.

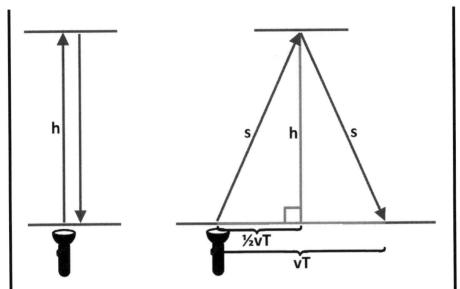

Note that the light beam does not look entirely vertical from Sarah's perspective, but is at an angle. This is because the train is moving, and so by the time the light reaches the mirror, the mirror is no longer directly above where the flashlight was when the beam was emitted. And when the beam returns back to the position of the flashlight, the train has moved a distance of vT. From Sarah's point of view, the total distance the light has traveled is not $2h$, but is $2s$ – a slightly greater distance. The time she records for the beam to return is this distance ($2s$) divided by the speed of light (c).

$$T = 2s/c$$

We can see from geometry that s is greater than h. So we already know that Sarah will record a different time for this experiment than Michael does! Her time will be greater than his, which means that Michael's moving clock ticks slower than her stationary clock. *Motion slows the rate at which clocks tick!*

But in order to see how much time is slowed by motion, we must get the equation that relates T to T'. This we accomplish by eliminating the variables s, and then h, as follows. From geometry, we know that a line segment representing the height of the train (h) is at a right angle to the floor of the train. This forms a right triangle with segment s as the hypotenuse. From the Pythagorean Theorem, we know that the sum of the squares of the legs of a right triangle is equal to the square of the hypotenuse. So we have:

$$\tfrac{1}{4}v^2T^2 + h^2 = s^2$$

And previously we found that Sarah measures the time of the experiment as: $T = 2s/c$, which solving for s we find: $s = \tfrac{1}{2}Tc$. Substituting this value for s into the previous equation we find:

$$\tfrac{1}{4}v^2T^2 + h^2 = \tfrac{1}{4}T^2c^2$$

Solving this equation for T we find:

$$T = \frac{2h}{\sqrt{c^2 - v^2}}$$

And we found previously that the time Michael measures for the experiment is: $T' = 2h/c$, which solving for h gives: $h = \tfrac{1}{2}T'c$. Substituting this value of h into the previous equation gives:

$$T = \frac{T'c}{\sqrt{c^2 - v^2}}$$

And simplifying this reveals the relationship between Michael's clock rate and Sarah's: the time dilation formula.

$$T = \frac{T'}{\sqrt{1 - \dfrac{v^2}{c^2}}} \tag{4.1}$$

The equation we just derived expresses the rate at which Sarah's clock ticks compared to the rate at which Michael's clock ticks, when Michael is moving at speed v relative to Sarah. The first thing we note is that there is indeed a difference. Unless Michael is stationary (in which case $v = 0$) there is a difference between the time he measures for the light-bounce experiment, and the time that Sarah measures for the same experiment. Thus, *velocity does affect the passage of time – the rate at which clocks tick.*

The second thing we note is that Michael's clock ticks *slower* than Sarah's. The train is moving at some speed less than that of light, so $v < c$. That means the denominator in the above equation is always less than one. Thus $T \geq T'$. As one example, if the denominator were ½, then Michael's clock measures one second for an event where Sarah's clock measures 2 seconds for the same event.   Clearly, Michael's clock ticks slower than Sarah's clock does. His passage of time becomes "stretched out" or *dilated*. For that reason, this effect is called *time dilation*.

The third thing we note is that this thought experiment did not specify what type of clock is needed. This indicates that the specific mechanism of measuring time does not matter. *Whatever* we use to measure time, that device will "tick" more slowly than its stationary counterpart does. It would not matter if Michael used a Rolex, or a Pendulum clock, or simply counted "one-Mississippi, two-Mississippi." All clocks are slowed by exactly the same amount.

As a result of the fact that all clocks are slowed equally, *Michael would not notice that his clocks are ticking slow.* This is a really important point, so don't miss it. Not only is Michael's stopwatch ticking slower than Sarah's, but Michael's other clocks are also slowed by the same amount. So when he compares two of his clocks, they both give the same (incorrect from Sarah's perspective) time. Furthermore, his pulse and brain are slowed by the same amount. So from his point of view, the passage of time inside the train *would seem perfectly normal.* From Sarah's perspective, Michael and any clock he has with him would be ticking very slowly; but from Michael's perspective everything inside the train would appear normal. You cannot "feel" or otherwise perceive the effect of time dilation on yourself.

Fourth, we note from the time dilation formula that the effect of time dilation is very, very small at the speeds at which we normally travel. To see how this works, let us substitute some velocity values into the time dilation formula to see how much 1 second of Michael's time (T') is stretched to Sarah's time (T).

If Michael's train travels at 65 miles per hour, this is equivalent to $v = 0.018056$ miles per second.   Putting this into the formula with $c = 186,282$ mi/s, means that 1 second on Michael's clock corresponds to 1.000000000000005 seconds on Sarah's clock. The difference is much less than a billionth of a percent – an entirely unnoticeable difference. It is therefore no surprise that we don't readily see the effects of dilation in our everyday lives; the effect is so small it would take an atomic clock to measure it!

Suppose Michael's train traveled at the speed of sound: about 760 miles per hour at sea level. This works out to $v = 0.21$ miles per second, which when inserted into the time dilation formula reveals that one second of Michael's clock is stretched to 1.0000000000006 seconds on Sarah's clock. Again, the effect is practically negligible.

Astronauts in Earth orbit travel at about 17,000 miles per hour. That converts to 4.72 miles per second. How much is their time dilated? Substituting $v = 4.72$ mi/s into the formula reveals that one second of an astronaut's clock is stretched to 1.0000000003 seconds of Earth time. Again, the effect would not be noticeable. This is perhaps why we find time dilation so counterintuitive. No one has ever really perceived it.

So how fast would Michael have to travel for the effects of time dilation to become noticeable to Sarah? The time dilation formula reveals that in order for Michael's clock to be a mere 1% slower than Sarah's, he would have to travel at a whopping 14% the speed of light. That works out to 94 million miles per hour! A train travelling that fast could circle the entire planet Earth in only 1 second! And that's the speed necessary for just a *one-percent* effect of time dilation. So if you had high hopes of extending your lifespan significantly through time dilation by traveling at high speed, you can forget it. Even our fastest spacecraft are nowhere near that kind of speed.

But what if we could reach such speeds? If we could build spacecraft that travel at a substantial fraction of the speed of light, how much time dilation would take place? The time dilation formula reveals the answer. The table below shows the amount of time dilation that takes place for a few selected velocities. The top row shows the speed of the spaceship compared to the speed of light. The bottom row shows how much 1 second of ship time would be stretched relative to Earth time.

v/c	0.01	0.1	0.14	0.333	0.50	0.75
T/T′	1.00005	1.005	1.01	1.061	1.15	1.52

v/c	0.80	0.9	0.99	0.999	0.9999
T/T′	1.667	2.29	7.09	22.37	70.71

The table reveals that time dilation is miniscule for velocities that are significantly less than the speed of light. Even at half the speed of light, a clock is only slowed by fifteen percent. However, as velocity gets closer to the speed of light the effect of time dilation becomes very substantial.

At ninety percent the speed of light, moving clocks are more than twice as slow as their stationary counterparts. And a clock moving at 99.99% the speed of light will tick only one second for every 70.71 seconds on Earth!

What happens when we let the velocity of our spaceship reach the speed of light? The denominator of the time dilation equation goes to zero, which is algebraically disallowed. This at least suggests that it may not be possible for a spaceship to actually reach the speed of light – a topic we will revisit in a later chapter. But, from a calculus perspective, we can say that in the limit as velocity approaches the speed of light, the effect of time dilation approaches infinity. This suggests that if you could travel at exactly the speed of light, your clocks would slow to the point that they would actually stop ticking completely. From light's point of view, every trip is instantaneous.

What Time Dilation Does **Not** Mean

There is a common misconception that often arises in discussions of time dilation. This is the idea that Michael's clocks only *appear* to tick slower from Sarah's perspective due to increasing time lag of the light – an optical effect. But time dilation is not merely an apparent slowing of clocks. Michael's clock is genuinely ticking slower than Sarah's clock from her perspective as we demonstrated in the above proof.

This can be confusing because there is indeed an optical effect – *in addition to genuine time dilation* – that can make Michael's clock *appear* to Sarah to be ticking slower than it really is. Suppose Michael's train is moving directly away from Sarah at half the speed of light. Sarah watches Michael's clock through a powerful telescope. Even if there were no time dilation, Michael's clock would still *appear* to be ticking slowly. This is because the distance between Michael and Sarah is increasing with each tick, so the light takes progressively longer to get to Sarah.

Suppose that when Michael's clock strikes noon, light from it takes one second to get to Sarah. Ignoring time dilation, when Michael's clock reaches 12:00:01, the light now takes 1.5 seconds to get to Sarah because Michael is now at a greater distance due to the high speed of his train. So she sees the second tick happen 1.5 seconds after the first, even though it really happened only one second after the first. As the distance continues to increase, so does the time delay. So Sarah will continue to see Michael's clock tick only 1 second for every 1.5 seconds that she experiences, due to the light-travel delay caused by Michael's increasing distance. This is *not* genuine time dilation; it is merely an optical illusion of sorts.

Unlike genuine time dilation, this optical effect is reversed if Michael is moving directly *toward* Sarah. In this case, the distance between Michael and Sarah *decreases* between each successive tick of the clock. Suppose that when Michael's clock strikes noon, it takes only two seconds for the image to reach Sarah. But when Michael's clock ticks to 12:00:01, it now takes only 1.5 seconds for this image to reach Sarah because the distance between them has decreased. So she would see Michael's clock ticking *twice as fast* as hers. This again is not genuine time dilation, but merely an optical effect – one that we will deal with in the chapter on the Doppler effect.

Genuine time dilation is not about what Sarah would see in her telescope, but rather what is actually taking place as computed from her reference frame. Recall that the geometric proof above did not involve what Sarah would see from a telescope, but rather what must actually be taking place inside the train according to the postulates of relativity. Moving clocks always tick slower than stationary ones due to time dilation, regardless of whether they are approaching or receding from the other observer. However, the optical effect of increasing or decreasing time lag is direction-dependent.

Therefore, what Sarah would actually see in her telescope will be a combination of actual time dilation, *and* any optical effect due to the increasing or decreasing time lag of light. Imagine that Michael was moving in a giant circle with Sarah at the center. In this case, there would be no optical effect because the distance between Michael and Sarah is constant. Therefore, in this case Sarah would see only the effects of genuine time dilation. Watching Michael through her telescope, she would see his clock ticking slower than hers.

Another misconception about relativity is that motion only affects the rate of clocks, not the passage of time. The claim is, "You shouldn't say 'time dilation'; you should say 'clock-retardation.'" But this seems to be a distinction without a difference. What, after all, do clocks measure but the passage of time? I could understand the above claim if only certain types of clocks were affected by motion. But this isn't that case. All clocks are affected equally, indicating that the flow of time itself has been stretched. From the proof above, we found that the *actual time* it took to do the experiment differs between Sarah's perspective and Michael's view. It does not matter what kind of clocks they use to measure time. The flow of time itself is affected by motion in such a way that the speed of light is constant in all inertial reference frames. The term "time dilation" is therefore perfectly appropriate.

Time Travel for Fun and Profit

The implications of time dilation are Earth-shattering. Granted, we do not (at present) possess the technology to accelerate spaceships to anywhere near the speed of light. But if we could, passengers on such ships would experience a high degree of time dilation. From the time dilation formula above (or from the table of sample values), we see that traveling at 99.99% the speed of light slows the passage of time by a factor of 70. So if we sent a spaceship with passengers at this speed to a distant star, what would happen?

Let's suppose we decide to visit a star that is 70 light-years away (400 trillion miles). A light-year is a unit of distance, equal to about 5.88 trillion miles. It takes light one year to travel this distance – hence the name. Travelling at 99.99% the speed of light, it would take our spaceship 70 years to reach the star, and then 70 years to return, for a total time of 140 years. That is 140 years as *measured by Earth clocks*. But the passengers aboard the ship would experience only two years during this time! They would reach the star in only one year as measured by their onboard clock. And they would take one year to return to Earth. Their passage of time has been slowed by a factor of 70, due to their high speed.

Of course, passengers on board the ship would not "feel" or notice the effect of time dilation because all things are slowed equally, including their perception of time. Their clocks, their pulse, their brain, and the aging of their body are all slowed by a factor of 70 relative to Earth clocks. For them, the entire trip really only takes two years. However, when they return to Earth, they would find that all of their friends and family on Earth have been dead for decades. They would return to an Earth that is 140 years older than the one they left only two years ago by their clocks.

It sounds like science fiction. And since we don't have spacecraft capable of traveling at such speeds, the scenario remains fiction for the present time. But the physics is real. From the geometric proof we saw above, time dilation must occur if the postulates of relativity are true as they appear to be. So the scenario of travelling to a distant star and back in very little time is entirely possible physically, though it is currently unattainable technologically. We will revisit this scenario in detail in a later chapter.

Experimental Confirmation

Einstein's discovery of time dilation was based on a "thought experiment" very similar to the one we used in this chapter. It is a proof by logical and mathematical deduction, one that needs no experimental support. Nonetheless, since the time of Einstein, there have been experimental confirmations of time dilation. But since the effect of time dilation is so miniscule at "normal" speeds, and since we lack the technology to move any reasonably massive object to a high enough speed for the effects of time dilation to become noticeable, our ability to experimentally confirm time dilation is limited. Yet, there are two ways in which we can experimentally confirm time dilation.

First, we can use very accurate and precise clocks that are able to measure even the infinitesimal effects of time dilation that happen at "normal" speeds. This has been done. In 1971 the Hafele-Keating experiment confirmed the effects of relativistic time dilation using atomic clocks that were flown on aircraft and comparing them with atomic clocks on the ground. The moving clocks did indeed experience time dilation consistent with the time dilation formula we derived above.[8]

Second, although we cannot accelerate planes, trains, or spaceships to any reasonable fraction of the speed of light, we can accelerate low-mass charged particles to speeds very near the speed of light. Some of these particles are known to "decay" (to change into a different particle and release energy) at a given rate. For example, the muon is a negatively-charged particle with a mean lifetime of 2.2 microseconds. And so it acts like a clock that automatically self-destructs after 2.2 seconds. What happens when we accelerate muons to nearly the speed of light? They experience time dilation and therefore live much longer than 2.2 microseconds as measured by Earth clocks.

This has been experimentally confirmed. Muons in particle colliders have been accelerated to 99.94% the speed of light. From the time dilation formula, we see that they should age 29 times slower than stationary muons. The mean lifetime of these high-speed muons has been measured to be 64 microseconds. That is, of course, 29 times longer than the lifetime of their stationary counterparts – a spectacular confirmation of the theory of relativity.

---

[8] The Hafele-Keating experiment also confirmed the effects of gravitational time dilation. This relativistic effect is addressed in a later chapter.

We have seen that motion affects the passage of time. We have also seen from the "painting the hallway" thought experiment that motion does *not* affect widths or heights. But one question remains: does motion affect lengths? That is, would Sarah and Michael agree on the length of the train? The answer may surprise you.

Chapter 5

Length Contraction

In the earlier chapters of this book, we found that the speed of light is always c (186,282 miles per second) faster than any inertial observer, regardless of how fast that observer is moving. It seems impossible that different observers would measure the same speed relative to themselves. But the resolution to this paradox is to realize that either lengths or times or both are not absolute, but observer-dependent. That is, motion affects either lengths, or the passage of time, or both. And we saw in the previous chapter that motion does affect the passage of time. But it remains to be seen if lengths are also affected.

As before, we imagine a train moving at high speed along the x-axis. Michael is a passenger on the train. From his perspective, the train is stationary and the Earth's surface is in motion. Sarah is standing by the track and is stationary relative to the Earth's surface. We found previously from the "painting the hallway" thought experiment that Sarah and Michael agree on the width and height of the train and the width and height of everything else. Motion along the x-axis does not affect measurements in the y- or z-axis.

But we found that time is affected; Michael's clocks run slower than Sarah's by an amount given by the time dilation formula which we derived in the previous chapter. We will use these findings along with the two postulates of relativity to see if lengths are affected or unaffected by motion.

Let L be the length of the train as measured by Sarah's tape measure, and let L' be the length as measured by Michael's tape measure. (If it turns out that $L = L'$, then we will conclude that lengths are not affected by motion.) As before, let v represent the velocity of the train relative to Sarah. As before, we will use a flashlight, a mirror, and a stopwatch to measure how long it takes light to travel to the mirror and back. But this time, we will orient the experiment along the x-axis. Michael will stand at the back of the train with his flashlight pointed forward in the direction the train is travelling. The mirror will be at the front of the train.

Michael will turn on the flashlight and start his stopwatch. The light will travel to the front of the train, bounce off the mirror, and back to Michael – at which point he will hit "stop" on the stopwatch and will record the time: T'. Sarah will watch from the tracks, and will record the time (T) from her perspective. We know from the previous chapter that Sarah will not record the same time as Michael, because Michael's clock ticks slower than hers. Nonetheless, this scenario will allow us to determine if lengths are affected by motion. As before, we will analyze the experiment first from Michael's reference frame and then from Sarah's frame. And we will derive the formula that relates L and L'.

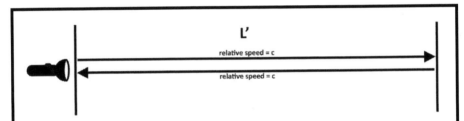

From Michael's reference frame, the light travels at speed c faster than he does. It travels to the front of the train (a distance of L') and then back again, as in the figure. So the light travels a total distance of $2L'$ and it does so at speed c and in time T'. Since distance is equal to speed multiplied by time, we have:

$$2L' = cT'$$

There are a few different ways we could analyze the situation from Sarah's perspective. The easiest scenario from a mathematical perspective is for Sarah to measure all distances *relative to the train*, while still recognizing that the speed of light is c *relative to her*. She will still use her tape measure and her clocks, but the position of the light beam will always be relative to the train. Under this scenario, the light, when moving toward the mirror, travels at a speed of $c-v$ faster than the train, as illustrated in the figure. This must be true since light travels at speed c relative to Sarah, and the train is moving at speed *v*. The Galilean velocity theorem does apply in this scenario since all the measurements are done by Sarah.

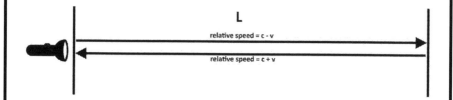

Once the light strikes the mirror, it reverses direction and is now moving in the opposite direction as the train. So from Sarah's perspective the light is moving at a speed *c+v* toward the Michael. Clearly, from Sarah's point of view, the light will take much longer to go from the flashlight to the mirror than it will to zip back. The time it takes for the light to go from the flashlight to the mirror will be *L/(c-v)*. And the time it takes to return will be *L/(c+v)*. The total time for this experiment will therefore be the sum of these two times:

$$T = \frac{L}{c - v} + \frac{L}{c + v}$$

All that remains is algebra. We simplify this equation by multiplying the first term on the right-hand side by *(c+v)/(c+v)* and the second term by *(c-v)/(c-v)* to combine these terms over a single denominator. This gives us:

$$T = \frac{L(c + v) + L(c - v)}{c^2 - v^2}$$

This further simplifies to:

$$T = \frac{2L/c}{1 - \dfrac{v^2}{c^2}}$$

We substitute this value of *T* into the time dilation formula (4.1) we derived in the previous chapter:

$$\frac{2L/c}{1-\dfrac{v^2}{c^2}} = \frac{T'}{\sqrt{1-\dfrac{v^2}{c^2}}}$$

And from the first equation above derived from Michael's reference frame, we know that $2L' = cT'$. Which means $T' = 2L'/c$. Substituting this value for T' into the previous equation we get:

$$\frac{2L/c}{1-\dfrac{v^2}{c^2}} = \frac{2L'/c}{\sqrt{1-\dfrac{v^2}{c^2}}}$$

We divide both sides by $2/c$ and multiply both sides by $1-v^2/c^2$ to obtain the following equation: the length contraction formula:

$$L = L'\sqrt{1-\frac{v^2}{c^2}} \qquad (5.1)$$

We first note that for any positive velocity, the length Sarah measures for the train (L) is not the same as the length that Michael measures for the train (L'). Since the velocity v, is some value less than that of the speed of light, the expression under the radical will always be something less than one. Therefore, L' will always be somewhat larger than L. What does this mean?

In general, whatever length Michael measures for the train, Sarah will measure a shorter length. Therefore, Michael's tape measure must be reduced in length compared to Sarah's tape measure, in order for him to estimate a greater value for a given length. The above formula shows that lengths contract in the direction of motion for any moving object! Hence, physicists refer to this effect as *length contraction*.

Second, we note (as we noted with time dilation) that the effect is extremely miniscule for "normal" speeds. Consider a car that is exactly ten feet long when stationary. The length contraction formula shows that when this car is moving at 65 miles per hour, it will compress to 9.9999999999999953 feet. No tape measure in the world could mark such a small change. This explains why we find length contraction so counterintuitive. In our everyday experience, we could not possibly notice any length contraction for moving objects; the effect is too small.

How fast would something have to travel for us to notice any appreciable length contraction? The length contraction formula shows that an object would have to travel at 14% the speed of light in order for it to contract by a mere 1% from its stationary length. If that number (14%) seems familiar, it should. This is the same speed at which time dilation grows to a 1% effect.

That brings us to our third observation. The formula for length contraction is very similar to the formula for time dilation. Both contain this term: $\sqrt{1 - \frac{v^2}{c^2}}$. But there is an important difference. The time dilation formula divides T' by this term, whereas the length contraction formula multiplies L' by this term. Since this term is always less than one for any positive velocity, this means moving rulers are always shorter than stationary ones, and moving clocks are always slower than stationary ones. Putting it another way, length is compressed due to motion, whereas time is stretched. *Motion has the opposite effect on length that it has on time.* Yet, the magnitude of the effect is identical. This is significant, and will lead to other discoveries later on.

The table below contains the amount of length contraction that occurs at various selected speeds. Specifically, it shows the length in feet that a one-foot long (when stationary) object would have at these speeds. As with time dilation, we note that the effect of length contraction is virtually negligible at low speeds, and only becomes significant when an object reaches a significant fraction of the speed of light. So it is not at all surprising that we don't notice any significant length contraction of objects that we ordinarily encounter. They are far too slow for the effect to be measurable.

v/c	0.01	0.1	0.14	0.333	0.50	0.75
L/L'	0.99995	0.995	0.99	0.943	0.87	0.66

v/c	0.75	0.80	0.9	0.99	0.999	0.9999
L/L'	0.66	0.60	0.44	0.14	0.045	0.0141

In the limit as its velocity approaches the speed of light, the length of any object (in the direction of motion) will approach zero. This again seems to hint that it may not be possible to accelerate an object up to the speed of light. If you could, the object would completely flatten and have no length at all.

As with time dilation, it is crucial to note that Michael would not perceive his own length contraction. The reason is that everything contracts by exactly the same factor. If the train contracts to half its stationary length, Michael will not notice because his tape measure has also contracted by half. So he will measure the same length for the train as he did when it wasn't moving. His body will contract by half, so he'll be able to make the same number of paces to the front of the train as when it was stationary. Nor would anything "look" strange, because even Michael's eyes will be length-contracted by the same amount. To Sarah, however, the train and everything within it would contract in the direction of motion, and would appear "squished."

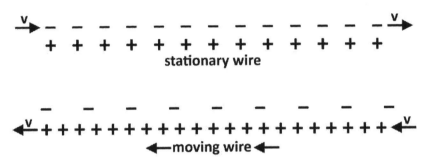

Length contraction is more difficult to verify experimentally than time dilation, at least directly. Nonetheless, experimental procedures are consistent with length contraction. For example, consider a stationary wire through which electricity is flowing. The electrons are moving along the wire from West to East with a mean velocity v. Suppose the wire is exactly electrically neutral, since the number of protons per unit length is equal to the number of electrons per unit length. Now, imagine that we move the same wire from East to West at speed v. The electrons are now stationary; so whatever length contraction they had when they were moving is now gone. Therefore, the electrons will be spaced slightly farther apart than they are in the rest frame. The protons, however, are now in motion. The protons will consequently experience length contraction; their spacing will decrease. As a result, any given length of wire will now contain more protons than electrons. The wire will have a net positive charge when moving. This does in fact occur, and is evidence of length contraction.

The Lorentz Factor

From the thought experiments in this chapter and the previous chapter, we have seen that the formula for time dilation is similar to the formula for length contraction. Both contain an expression of the square root of the quantity of one minus the velocity squared divided by the speed of light squared. This mathematical expression occurs so frequently in the physics of relativity that it is convenient to assign a symbol to its reciprocal. Physicists use the lowercase Greek letter gamma (γ) to denote this expression, which is called the Lorentz factor:

$$\gamma = \frac{1}{\sqrt{1 - \dfrac{v^2}{c^2}}} \qquad (5.2)$$

Note that the Lorentz factor is the *reciprocal* of the expression in the square root. From this definition, we can write the time dilation formula and length contraction formula very succinctly as follows:

$$L = {L'}/{\gamma} \qquad (5.3)$$

$$T = T'\gamma \qquad (5.4)$$

Some textbook authors prefer to write the primed quantities in terms of the stationary ones. To get this result, we simply divide each equation above by gamma. So we have $L' = L\gamma$ and $T' = T/\gamma$.

The figure shows a plot of the Lorentz factor (γ) versus speed (in terms of the fraction of the speed of light) up to 99% the speed of light. We have highlighted a few selected values of interest. The Lorentz factor has a minimum value of 1, which it attains when velocity is zero. The Lorentz factor is greater than 1 for all positive velocities, but is only larger than 1 by a miniscule amount for "ordinary" velocities. It only becomes large when velocity approaches the speed of light. As velocity approaches the speed of light, the Lorentz factor approaches infinity.

We can think of the Lorentz factor as the "strangeness factor" of relativity. At the speeds we normally experience in our everyday lives, the Lorentz factor is so close to 1 that the difference isn't measurable. So we don't notice any weird effects like time dilation or length contraction. For the Lorentz factor to reach a value only 1 percent above unity ($\gamma = 1.01$), the velocity must reach 14% the speed of light. Since virtually nothing in our everyday experience moves anywhere near this velocity, the strange effects of relativity are always less than one percent.

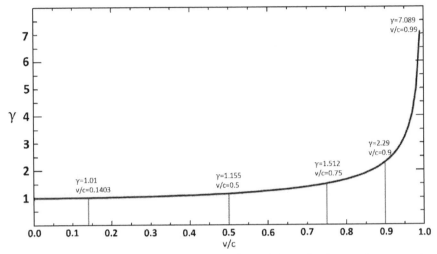

To calculate how much time is slowed by velocity, or how much lengths are contracted, we need only multiply or divide respectively by the Lorentz factor. To decide quickly whether we need to multiply or divide by γ, the following two statements are helpful: (1) *Moving clocks tick slow.* (2) *Moving rulers are short.* Remember that one second of a moving clock is "stretched" (dilated) so that it lasts more than one second on a stationary clock. And a moving ruler will be contracted in the direction of motion so that it is shorter than a stationary one of identical construction.

We have found that velocity affects both the passage of time and the length of any moving object. But what is even more surprising is the remarkable way in which these two effects work together. They are not the separate effects they may seem on the surface. Rather they are inexorably linked, like two sides of one coin. In the next chapter, we will begin to see how space and time are linked in a most surprising way.

Chapter 6

The Twin Paradox

As a young boy, Luke dreamed of becoming an astronaut. He would often imagine wonderful adventures on distant worlds. What would it be like to set foot on the rocky terrain of some alien planet? As a child of the late twenty-third century, Luke was well acquainted with the history of manned space exploration. He had read of the first men to set foot on the moon over three centuries earlier. And yet even by Luke's time, human space travel was still quite limited. No human had yet ventured beyond the solar system, and even the Mars colony was only in its infancy.

So it was with great interest that Luke began studying the recent advancements in anti-matter propulsion. These new technologies would allow spaceships to travel at velocities that were previously unattainable. By Luke's 18th birthday, the Cygnus Space Program was already well underway, and had begun recruiting potential astronauts for the first manned deep-space mission. After a seven-year period of rigorous intellectual and athletic training, Luke was selected to be an astronaut for the first manned mission to visit another star. The Cygnus spacecraft would be capable of reaching a speed of 99.99% the speed of light. Its destination was a planet orbiting the distant star 16 Cygni – a star 70 light-years (411 trillion miles) away from Earth. Yet, the wonderful effects of relativity would allow Luke to travel this distance in far less than 70 years by his clocks.

The story is fictional of course. Yet, the physics is real. The principles and equations we derived in the previous chapters inform us of exactly what Luke would experience on his journey, even though human beings do not yet possess the technology to implement it. What we find particularly interesting is not just Luke's experience in the spacecraft, but the experience of his twin sister Carrie, who remains on Earth.   Her explanation of how things happen will differ somewhat from Luke's description.   And yet, everyone will agree without contradiction on the objectively measurable result.

To simplify the math a bit for our purposes, let's suppose that the Cygnus spacecraft travels from Earth to 16 Cygni at a constant speed of 99.99% the speed of light (relative to Earth).[9]   We compute the time dilation and length contraction of the ship, from the formulas in the previous two chapters.   The easiest way to do this is to first calculate the Lorentz factor (γ) from equation 5.2, and then use the shorthand formulas we discussed in the previous chapter (equations 5.3 and 5.4).   The velocity compared to the speed of light is $v/c = 0.9999$, and from the previous chapter, we found:

$$\gamma = \frac{1}{\sqrt{1 - \dfrac{v^2}{c^2}}}$$

Substituting our value for v/c we find that in this case $\gamma = 70.712$, which means Luke's clocks will tick just about seventy times slower than Carrie's clocks.   Moreover, his ship will be compressed in the direction of motion so that the ship and everything within it are 70 times shorter than they were when stationary.   Furthermore, we conveniently selected a star that is 70 light-years away.   So at 99.99% the speed of light, it will take just over seventy years for Luke to reach 16 Cygni, as measured by Carrie.

But Luke is well-trained in relativistic physics.   He knows that at 99.99% the speed of light his time will slow by a factor of just over 70 compared to time on Earth. So from his perspective, the trip will take only one year.   He will age only one year, and experience only one year.   All clocks are slowed equally along with Luke's perception of time.   So everything appears normal to him.   Yet, the trip lasts only one year.   Both Luke and Carrie are 25 years old when the Cygnus departs.   But upon

---

[9] A more realistic scenario is for the ship to accelerate up to 99.99% during the first half of its journey, and then decelerate back to 0 for the last half of its journey.   This means that the Lorentz factor will constantly change, and makes computing the time dilation a bit more challenging.   It would place Luke in a non-inertial reference frame. For the purposes of illustrating the concept, the above simplification will suffice.

arrival, Luke will be 26 years old, whereas Carrie will be 95 years old. Remarkably, these twins have very different ages.

After a brief period of time exploring the 16 Cygni system, Luke begins his journey home. Again, his ship moves at a constant speed of 99.99% the speed of light. So the return trip takes an additional 70 years as measured by Earth's clocks. But from Luke's perspective, the trip lasts only one year. He returns to Earth, a 27-year old man. But the Earth he returns to has aged 140 years in his absence, and is now in the early twenty-fifth century. His twin sister has been dead for decades. (She would have been 165 years old.) But two of her grandsons are still living, though both are in their nineties. They are delighted to meet their 27-year old great uncle for the first time, and happily introduce him to several of their children and grandchildren.

From Carrie's perspective, Luke is able to travel 70 light-years while only aging one year due to time dilation. But how do we account for Luke's ability to make the trip in only one year from *his* point of view? Recall that from the first postulate of relativity, Luke is allowed to consider his spaceship to be stationary. From his point of view, the rest of the universe is moving at 99.99% the speed of light. But if Luke is stationary, then he is experiencing no time dilation. So how is he able to make a 70-light-year-trip in only one year? From his point of view, he is stationary and the star 16 Cygni is traveling toward him at 99.99% the speed of light. So how does the star reach him in only one year, as opposed to 70 years? How do we resolve this paradox?

The answer involves length contraction. As soon as Luke's ship jumps forward to 99.99% the speed of light, from his point of view his ship is stationary and the universe is moving backward at 99.99% the speed of light. Therefore, from Luke's point of view, *the universe will experience length contraction*. It contracts in the direction of motion by γ (the Lorentz factor), which we found earlier to be around 70. So from Luke's vantage point, the distance to 16 Cygni has been reduced by a factor of 70. The star is now only one light-year distant. And since the universe, including 16 Cygni, is moving toward Luke at 99.99% the speed of light, the star reaches the position of his ship in only one year.

Both twins agree that Luke will reach the star when he turns 26. But the explanation depends on the reference frame. From Carrie's perspective, the answer is time dilation of Luke. From Luke's perspective, the answer is length contraction of the universe. It will be tempting to ask, "Which answer is correct?" But of course, both are correct. Luke's explanation is correct from his reference frame. And Carrie's explanation is correct from her reference frame. That is why this branch of physics is called "relativity."

One other thing to note is that we now see the reason for what we discovered in the previous chapter. Recall, we found that lengths compress by exactly the same factor that time stretches – by the Lorentz factor (gamma). We now see why this must be the case. Both Luke and Carrie must agree that Luke ages only one year when he arrives at 16 Cygni – this is objectively measurable because they could both look at the time on his clock when he arrives (Carrie looking at it through a powerful telescope). From Carrie's perspective, this is because Luke's time is slowed by a factor of 70. But from Luke's perspective it is because the universe compressed by a factor of 70. The factor must be the same in order for them to record the same objectively measurable result.

Application

There is a real-world verification of the twin paradox. In a previous chapter, we discussed the fact that muons – charged particles with a mean lifetime of 2.2 microseconds when at rest – are time-dilated at high speed. Muons are produced when cosmic rays strike Earth's atmosphere, several miles above the ground. These muons have a typical speed of around 99.5% the speed of light. But in 2.2 microseconds, a particle traveling at 99.5% the speed of light can travel less than half a mile. So apart from relativity, we wouldn't expect many muons to reach Earth's surface. But they do. The reason is that at 99.5% the speed of light, the Lorentz factor is around ten. Thus, the muons' time is stretched by a factor of ten, allowing them to persist for over 20 microseconds as measured by Earth clocks. This is long enough for many of them to reach the surface.

But consider the muon's reference frame. From its point of view, it is stationary and it is the Earth that is moving upward at 99.5% the speed of light. But if the muon is stationary from its own perspective, it is not experiencing any time dilation. So how is it able to reach the surface of Earth? The answer is length contraction. From the muon's perspective, the distance from its origin in the atmosphere to the surface of Earth has contracted by a factor of ten. Since the distance is reduced, the muon is able to reach the surface in less than 2.2 microseconds from its perspective.

The Counterintuitive Nature of the Resolution

We tend to think that the length of something is absolute: either a ruler is a foot long or it isn't. But from the previous chapter we found that this isn't so. Length depends on the observer's state of motion. A ruler can be one-foot long from one reference frame, and yet *not* a foot long from a different reference frame. Moreover, we tend to think of time as absolute: either something takes one minute or it doesn't. But in reality, something can take one minute as measured in one reference frame, and not one minute in another reference frame.

Some may ask, "Doesn't this violate the law of non-contradiction?" The law of non-contradiction is a fundamental principle of logic. It states that a proposition and its negation cannot both be true at the same time and in the same sense. Or more colloquially: "You can't have A and not-A at the same time and in the same sense" where "A" is any truth claim. The last part of this definition is crucial: "in the same sense." It would indeed be contradictory to say that a ruler is one-foot long from Luke's reference frame, and it is not one-foot long from Luke's reference frame at the same time. But there is nothing in logic that requires lengths to be the same in *all* frames of reference.

But doesn't it seem like they should be? Our natural inclination is to be deeply skeptical of time dilation and length contraction. Such concepts go against our intuition. But there is a reason for this. Intuition is based on experience. And most human beings have no experience traveling at speeds at which time dilation and length contraction become noticeable. Everything in our experience seems to confirm that velocity has no effect on the flow of time or the length of any object. But our experiences with velocity are very limited. Thus, our intuition serves us well at low velocities because time dilation and length contraction may be neglected. But our intuition fails in areas in which we have little or no experience.

We would do well to remember that God is under no obligation to make the universe conform to our expectations. To a very young child, it is counterintuitive that the Earth is a sphere. All of his (limited) experiences seem to show that the Earth is flat. And indeed, on very small scales, the Earth is pretty close to flat. But as the child grows and gains experience by flying on a plane where the curvature of the Earth starts to become visible, he comes to realize that the Earth is round after all.

My point here is that we should never deny a scientific discovery on the basis that it is counterintuitive, particularly when such discoveries pertain to an aspect of creation with which we have little or no experience. Our intuition serves us well in areas of daily familiarity. But it is fallible, and based on limited experiences. As our circle of experiences enlarges, our intuition will gradually adjust to accommodate new data. Areas of thought that were once counterintuitive will become intuitive as our experience in those areas grows.

The theory of relativity is no exception to this. Our natural inclination to think that time and space are absolute and unaffected by motion is simply wrong. The real universe just isn't that way, and this is objectively provable. It is helpful to keep this in mind because we are about to explore some aspects of relativity that are even more counterintuitive than those we have already seen. Nonetheless, these aspects are just as provable as those we have covered so far. It is quite wonderful that God has designed a universe that works perfectly with no genuine contradictions and that is nonetheless very counterintuitive to us. It humbles the human mind.

Multiple Paradoxes

The twin paradox described above really consists of several paradoxes, only some of which we have covered so far. It is a paradox that twins could end up with very different ages. We've already seen the resolution to this paradox: time dilation. Time does not flow at the same rate in all reference frames, and therefore twins can age very differently depending on their motion. Another paradox is how Luke is able to travel to 16 Cygni in only one year from his reference frame – in which he is stationary. We've seen that length contraction is the answer. It is counterintuitive that what is the result of time dilation from one reference frame is the result of length contraction in a different reference frame. This suggests that space and time are really merely two sides of the same coin; and which side someone sees will depend on his or her state of motion.

There are three remaining paradoxes in this scenario. The first of these concerns an apparent lack of symmetry in this situation. We know from the first postulate of relativity that the laws of physics work the same in all inertial reference frames. Carrie is allowed to consider herself stationary, and Luke travels at 99.99% the speed of light relative to her. From Carrie's perspective Luke experiences a strong time dilation effect, and ages only one year for every seventy on Earth. So when he returns, naturally, she would be much older than he.

But why isn't the situation symmetric and reversible? From Luke's point of view, he is stationary. It is the rest of the universe (including 16 Cygni, the Earth, and Carrie) that moves at 99.99% the speed of light. So from his point of view, it is the universe that should experience extraordinary time dilation, and age only one year for every seventy years by his clock. So when he meets up with Carrie again, why isn't he much older than she is, if *she* is the one that experienced the time dilation?

But there is a very important event that breaks the symmetry. Luke is the one that changes direction – not Carrie. For the first 70 years (by Earth clocks) he is moving *away* from the Earth at 99.99% the speed of light. But upon reaching 16 Cygni, he turns around and then approaches the Earth at 99.99% the speed of light. Luke experienced an enormous change in velocity. Why is this important? It's because when Luke changes direction *he is not in an <u>inertial</u> reference frame*.

Recall that an inertial reference frame is a non-rotating frame that moves at constant velocity. Only inertial observers are permitted to consider themselves stationary such that the laws of physics work the same as any other inertial frame. Accelerating frames are not inertial. When Luke is moving toward 16 Cygni at constant speed, he is inertial. He is allowed to consider himself stationary and therefore will not experience any time dilation from his point of view. This is also true as he returns to Earth at constant speed. But when he reaches 16 Cygni and turns around, he experiences a drastic change in velocity, slowing from 99.99% the speed of light, to zero, and then to 99.99% the speed of light in the opposite direction. During this change, Luke is not permitted to consider his frame stationary in terms of the laws of physics; he must admit that he is the one accelerating – not Carrie.

Relativity still allows us to calculate what will happen in a non-inertial reference frame. It is simply that such a frame may not be considered stationary. Luke will have to perform a *Lorentz transformation* to find out how his turning around has affected his flow of time relative to observers on Earth. We will discuss the details in a later chapter. For now, it is sufficient to recognize that from Luke's point of view, it is when he turns around that the Earth experiences an apparent jump forward in time. From his point of view, it is during this turnaround that Carrie will age around 140 years. Whereas, from Carrie's point of view, Luke's clock was slowed for the entire journey. But both would agree on the objectively measurable result: Luke returns to an Earth that is 140 years older than when he left, even though he himself has aged only 2 years.

The Time Dilation Paradox

There is an additional paradox. This again concerns the apparent symmetry between Luke's reference frame and Carrie's reference frame. On the outward journey, *before* Luke reaches 16 Cygni and begins to turn around, he is in an inertial reference frame, moving at a constant speed of 99.99% the speed of light relative to Carrie. Therefore, from Luke's reference frame, he considers himself to be stationary, and Carrie is moving at 99.99% the speed of light away from him. But from Carrie's point of view, Luke is moving, and so his clocks tick slow by a factor of 70. Now see if you can solve the following riddle:

> From Carrie's reference frame, Luke's clocks are ticking *slow*. Therefore, from Luke's reference frame Carrie's clocks must be ticking _____.

What is the answer? If you said, "fast" then you are in good company. Most people would answer this way, because it is the intuitive response. While intuition serves us well in everyday experience, time dilation is not an everyday experience. It certainly seems like "fast" would be the answer, but it isn't. Recall the time dilation formula we derived earlier, and the trend it implies: moving clocks tick *slow*. They are never fast. Motion only slows time; it does not speed up time.[10] From Luke's perspective, Carrie is in motion, and therefore her clock will be ticking slow relative to his. If you said, "slow" – congratulations! You are ahead of the curve in correctly understanding relativity! Allow me to elaborate on this (perhaps counterintuitive) answer.

From Carrie's point of view, Luke is in motion. And moving clocks tick slow. Therefore, Luke's clock ticks slow (slower than Carrie's clock). But from Luke's point of view, he is stationary and Carrie is in motion. And moving clocks tick slow. Therefore, Carrie's clock ticks slow (slower than Luke's clock). Each observer would measure the other clock ticking slower than his or her own clock! Now that is truly counterintuitive.

---

[10] This is true under the Einstein Synchrony Convention, but not necessarily under other conventions. These alternate synchrony conventions will be discussed in a later chapter.

The paradox arises from our expectation that there is such a thing as absolute time. So if Luke's clock is genuinely ticking slower than Carrie's clock, we deeply suspect that Carrie's clock must be genuinely ticking *faster* than Luke's clock. We want to ask, "Whose clock is *really* the slow one?" But any answer to this would violate the first postulate of relativity. Luke's reference frame and Carrie's reference frame are equally legitimate.

How can it be that Luke observes Carrie's clock ticking slower than his, whereas Carrie observes Luke's clock ticking slower than hers? Before we explore the answer, I need to make clear what the answer is *not*. The answer is not simply an optical illusion due to the increasing distance between Luke and Carrie. I mentioned this briefly in the time dilation chapter. But it's important to review this here.

It is not just that Carrie *sees* Luke's clock merely *appearing* to tick slower than hers because of the increasing time lag of light signals due to the increasing distance of Luke's ship. There is indeed such an optical effect. But both Carrie and Luke are smart enough to subtract this effect from their measurements of the other person's time. After subtracting this effect, each observer *still* computes that the other clock is ticking slower than his or her own clock. There is genuine time dilation in addition to an optical time lag.

The resolution to this paradox is to recognize that the only objectively measurable result is what Carrie and Luke each actually see – which includes both genuine time dilation and the optical effect of increasing time lag. They will disagree on the relative contribution of each effect. Let's pick some actual values to illustrate this.

We found that at 99.99% the speed of light Luke's clock is slowed 70.712 times relative to Carrie's clock. This means that between two consecutive ticks of Luke's clock, his ship has been traveling for 70.712 seconds by Carrie's clock. At 99.99% the speed of light, Luke's ship will travel a distance of 13,171,000 miles in that 70.712 seconds. Therefore, the light from each tick of the clock will have to travel a distance of 13,171,000 miles *more* than the light from the previous tick did. It takes light 70.705 seconds to traverse this distance. Therefore, Carrie will see in her telescope that each tick of Luke's clock happens 141.42 seconds after the previous tick.

So Luke's clock visually appears to Carrie to be 141.42 times slower than her clock. Roughly half of this apparent slowing (70.712) is due to actual time dilation. The other approximately half (70.705) is an optical effect due to the increasing distance the light must travel between consecutive ticks of Luke's clock. From symmetry, Luke would also see

Carrie's clock appear to tick 141.42 times slower than his own. He would conclude that about half of this is due to actual time dilation, and half is due to the increasing time lag of the light as the distance to Carrie increases.

Though Carrie and Luke disagree about whose clock is experiencing the time dilation, they both *agree on what the other person would see in their telescope*. That is, Carrie agrees that Luke should see her clock ticking 141.42 times slower than hers. However, the *cause* is different. Consider the following:

Suppose Luke passes by Carrie at time = 0. One second later, Carrie's clock has ticked one second. But how long will it take the image of that tick to catch up to Luke? In that one second, he has traveled a distance of 186,264 miles. But he continues to move away at 99.99% the speed of light. And so the light travels only 0.01% faster than Luke (from Carrie's perspective). To make up the distance (186,264 miles) at a relative speed of $c-v$ (0.0001c) will therefore be: 186,264/0.0001c = 9,999 seconds. And this second tick is one second after the first tick, so 9,999+1 = 10,000. Thus, Carrie estimates that Luke will see the image of the second tick of Carrie's clock 10,000 seconds after the first tick due to the time lag of the light beam. But Carrie knows that Luke's clock is slowed by a factor of 70.712 and therefore 10,000 seconds on her clock will register as only 10,000/70.712 seconds on his clock which is 141.42 seconds. That's exactly what Luke sees!

So there is no contradiction in terms of any objectively measurable quantities. Luke and Carrie both agree on what the other person should see at any given time. They only disagree on the relative contribution of genuine time dilation and the optical effect of time lag caused by the increasing distance between these observers. This is similar in essence to an earlier paradox: both Luke and Carrie agree that Luke ages only one year in the time it takes to reach 16 Cygni – but they disagree on whether the cause is due to time dilation of Luke's clock or length contraction of the universe. This is a common feature of the theory of relativity, and implies that space and time are not the independent entities that we might naively assume.

The "Trapping the Train" Paradox

An even stranger paradox arises when we consider length contraction. Let's go back to an earlier scenario, but with slight modification. Suppose that Michael is traveling on a train at 90% the speed of light relative to

Earth's surface. At this speed $\gamma = 2.29$, so lengths contract and time dilates by a little over a factor of two.

As before, Sarah is stationary relative to Earth's surface. Let's suppose that she is also in a train of identical construction to Michael's, but her train is stationary on a parallel track. Consider the moment when Michael's train passes by Sarah's train. From Sarah's point of view, Michael's train will be contracted in length by a factor of 2.29. So it will be a bit less than half the length of her train. Now, solve the following riddle:

"From Sarah's reference frame, Michael's train is 2.29 times shorter than her train. Therefore, from Michael's reference frame, Sarah's train is 2.29 times _____ than his train."

As before, our intuition suggests that "longer" is the right word to fill in the blank. And as before, our intuition is wrong. "Shorter" is the correct answer. From Michael's reference frame, he is stationary and Sarah is the one moving at 90% the speed of light. Therefore, *her* train will be length contracted – not his. We proved in the length contraction chapter that moving rulers are shorter than stationary ones. So how is it that each observer can see the other train shorter than his or her own train at the very same moment in time?

It seems like it should be possible to determine which train is "actually" shorter. Sarah proposes the following experiment. Along the train track, she finds a tunnel that is the exact length of her stationary train, which is also the length of Michael's train *when he is stationary* relative to Earth. This is illustrated in the above figure. Now, Sarah claims that Michael's train contracts when he is moving. So it will be shorter than the tunnel when it passes through, as shown in the top panel of the figure below. But from Michael's perspective, he is stationary and it is the *tunnel* that is moving, and therefore the tunnel contracts. So from his perspective, the tunnel will be shorter than this train at the moment he is inside it, as shown in the bottom panel of the figure below.

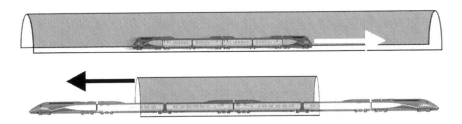

To see who is right, Sarah installs a door at each end of the tunnel. She intends to prove that, in fact, Michael's train is shorter than the tunnel by closing each door at the moment Michael's train is in the middle of the tunnel. She will thus temporarily trap Michael in the tunnel, just long enough to take a photograph, and then open both doors allowing the train to continue on its way. This will prove that indeed Michael's train is contracted since it easily fits inside the tunnel.

Michael is skeptical that Sarah can do this. From his point of view, the tunnel is length contracted, and thus shorter than his train. There is no way Sarah will be able to trap a long train in a short tunnel. But he agrees to the experiment.

It seems like this experiment should establish once and for all who is right. After all, if Sarah succeeds in temporarily trapping the train in the tunnel, then it will prove that Michael is the one who is actually experiencing length contraction. But if Sarah is unable to trap the train because it is too long, then it will prove that Michael is right and the tunnel is experiencing length contraction.

So what happens? Either result seems to violate the first postulate of relativity – that all inertial reference frames are equally legitimate. It seems that we are missing something. To understand the resolution to this paradox, we need to explore one more strange aspect of relativity in the next chapter – the relativity of simultaneity.

Chapter 7

The Relativity of Simultaneity

Suppose a scientist developed a new sophisticated computer model capable of predicting weather phenomena with unprecedented accuracy and precision. He claims that he can actually predict when and where lightning will strike! Of course, not many people believe him. This scientist predicts that two bolts of lightning will strike *simultaneously* (at exactly the same time) today at 12:00 noon. He predicts that one will strike at location A and the other at location B – a short distance away. Sarah decides to do an experiment to test his claim.

Sarah goes to location M – exactly half way between A and B. She places mirrors at a 45-degree angle relative to each of her eyes, so that one eye is looking at location A while the other eye looks at location B. If indeed the lightning bolt at A happens at the same instant as the lightning bolt at B, then she will see both flashes at exactly the same time. And she will record the time.

Granted, it takes a small amount of time for the light to go from A to M and from B to M. But since the distance is the same, the time delay will be identical since the speed of light is constant. Sarah can easily subtract this small time delay from when she observed the flashes to find the time when the flashes actually happened. Suppose she does all this and indeed sees both bolts happen at the same time, and after subtracting the light travel time, she computes that indeed both lightning strikes happened at 12:00 noon. She concludes that the scientific prediction is right. Or is it?

By divine providence, locations A and B just happen to lie along the track that Michael frequently travels in his high-velocity train. So Michael decides that he too will test the predictions of the scientist, since his train happens to be passing by at 12:00. Let's suppose that Michael is traveling at some high speed from left to right, and passes by Sarah at exactly 12:00; they are both essentially at the same location for just that instant. And that is the very instant that Sarah claims both lightning bolts strike at A and B.

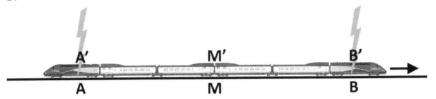

From Michael's perspective, the lightning bolts strike at A' and B' at the base of his train, which match A and B along the tracks at the moment of impact. We might imagine that they leave scorch marks at A' and B' on the train, and at A and B on the tracks, so there is no doubt where the lightning hit. Michael is in the middle of his train at M' exactly halfway between A' and B'. Will Michael agree with Sarah that the lightning strikes are simultaneous?

From Sarah's perspective, both lightning bolts strike at exactly noon. The light from each bolt takes a small amount of time to arrive at M. But while the light is traveling toward M, the train is moving to the right. Therefore, the light waves from the lightning bolt at B will reach Michael's eyes before the light waves from lightning bolt A as shown in the figure below.

Therefore, Michael will see the lightning bolt in front of the train before he sees the lightning bolt behind the train. Now from his perspective, he is stationary and the world is moving to the left at high speed. From his reference frame the speed of light is always exactly c relative to himself. And since he is exactly in between A' and B', the time lag from the light at A' will be the same as the light from B'. Therefore, Michael will conclude that lightning bolt B' actually happened first, and lightning bolt A' happened second. They are not simultaneous from his perspective. He calculates that lightning bolt B happened slightly before noon, and lightning bolt A happened slightly after noon.

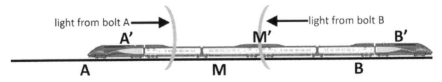

Note that both Sarah and Michael agree that Michael will see lightning bolt B before he sees bolt A. From Sarah's perspective, this is because Michael has moved toward B and therefore the time lag for the light from A is greater, even though she considers both lightning bolts to be simultaneous. But from Michael's perspective, he is stationary and exactly between A' and B'. From his point of view the time lag is equal, and thus the lightning bolt at B/B' must have actually happened before the lightning bolt at A/A'.

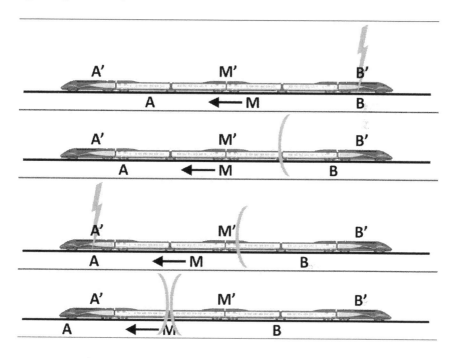

Furthermore, Michael will agree that Sarah sees the light from both lightning bolts at the same time. But from his perspective, this is *not* because the lightning strikes were actually simultaneous. Rather, it is because Sarah is moving to the left at high speed. And so the light from bolt B takes longer to "catch up" and reach her. Therefore, the actual lightning strike at B had to happen earlier than A in order for the light to reach Sarah at the same time as A. The accompanying figure illustrates this from Michael's reference frame.

The top panel is the first lightning strike, which from Michael's reference frame happens just a bit before noon. The second panel is noon – the moment Michael and Sarah are at the same location. The light from bolt B has traveled some distance, but has not yet reached either M or M'. In the third panel, the time is now shortly after noon, and lightning strikes at A/A'. In the fourth panel, light from both lightning bolts reaches Sarah simultaneously.

Like lengths of time and measurements of length, we have found from the above thought experiment that the concept of simultaneity is relative to reference frame. Thus, if one observer considers two events to be simultaneous, another observer in a different reference frame will not generally agree. What is simultaneous in one reference frame is not simultaneous in some other reference frame. This is a very important principle of relativity, and must be considered in any model that attempts to deal with distant starlight within the biblical timescale.

In the above thought experiment, we did not specify any quantities. And so we do not yet have a formula that specifies how much earlier lightning bolt B happens before A from Michael's perspective. But we have demonstrated the conceptual principle that "simultaneous" is dependent on reference frame. Furthermore, we have demonstrated the direction of the effect. That is, when Sarah observes the lightning strikes to be simultaneous, and Michael is moving from left to right, he will find the lightning bolt on the right to happen before the one on the left.

If we had reversed the direction of the experiment and had Michael moving right to left, then we would have found that he computes the left lightning bolt to happen before the right lightning bolt. So the moving observer will calculate *earlier times than the stationary observer for events in front of him, and will compute later times for events behind him.*

This itself is an interesting phenomenon, and we will revisit it in greater detail later. But for now, notice that the relativity of simultaneity implies that *the order in which some events occur can be observer-dependent!* Imagine the above experiment, but with *two* trains traveling in opposite directions on parallel tracks. Both pass point M at exactly noon. The train moving to the right will compute that lightning strike B happens before A. But the train moving to the left will compute that lightning strike A happens before B. This has profound implications for any discussion of time travel into the past, or faster-than-light communication.

Resolving the "Trapping the Train" Paradox

We are now able to resolve the paradox introduced in the last chapter. Recall that in the previous scenario, Michael's train is the same length as the tunnel when both are stationary. When Michael moves at high speed, Sarah notices that his train contracts, and is much shorter than the tunnel. But from Michael's reference frame, he is stationary and the tunnel is moving; so the tunnel contracts and is much shorter than his train. Sarah attempts to prove that Michael really is length contracted by closing both doors on the tunnel for an instant when his train is inside. But from Michael's reference frame, Sarah will not be able to trap his train since the tunnel is too short.

So is Sarah right? Is she able to close both doors on the tunnel at the same time, thereby temporarily trapping Michael's train within? Or is Michael right, and Sarah will be unable to close the doors on the tunnel at the same time since the train is longer than the tunnel from his perspective? The answer, of course, is *both*!

Did you catch the key phrase in the above paradox? The issue is whether or not Sarah can close both doors *at the same time*. But from the relativity of simultaneity we discovered above, we found that when one observer computes that two things happen at the same time (whether two bolts of lightning, or the closing of two doors), another observer in motion relative to the first will *not* find that these two things happened at the same time.

From Sarah's perspective, she will succeed in temporarily trapping Michael's train. She will close both doors at the same time while the train is entirely in the tunnel. But Michael is moving at high speed from left to right relative to Sarah. So from his perspective, the right side door closed and reopened first (top frame in the figure), then *at a later time* the left side door closed and reopened (bottom frame in the figure).

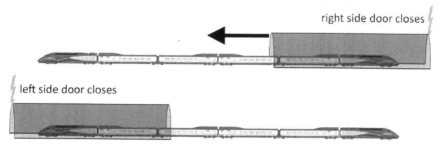

right side door closes

left side door closes

Sarah sees the experiment verify her reference frame, but Michael sees the same experiment verify his reference frame. Of course, the experiment must vindicate both reference frames, because the first postulate of relativity insists that all inertial frames are equally legitimate. Both Sarah and Michael are *right*.

But Sarah took a picture of the tunnel with both doors closed and the train within. Doesn't this prove that her reference frame is the true one? Michael would disagree. He points out that from his reference frame, Sarah and her camera were moving to the left at high speed. And so the light from the right side door took much longer to catch up with them than the light from the left side door. Sure, Sarah (and the camera) see *the images* of both doors closed at the same time. But from Michael's point of view, the light from the right-side door had to travel a greater distance to get to the camera, and therefore took longer. The photograph does not compensate for the time lag of the light. It is the same effect as Sarah seeing the images of two bolts of lightning happen at the same time, even though from Michael's perspective bolt B happened before bolt A.

The relativity of simultaneity also resolves another paradox: It explains why two observers both see the other's ruler shorter than their own. To measure the length of something, you must know the positions of both ends of it *at the same time*. But different observers will disagree on what constitutes "the same time." Michael will observe that Sarah's ruler is genuinely shorter than his own. But he will agree that Sarah will see his ruler as shorter than hers, because she doesn't record the endpoints at the same time, and the ruler has moved a bit. Both observers agree on the result that each observer sees his own ruler as the longer one. However, whether the cause is due to length contraction or the relativity of simultaneity depends on the reference frame.

In the next chapter, we will quantify the relativity of simultaneity and discover the equations by which we can relate any times or positions from one reference frame to any other reference frame.

Chapter 8

Coordinate Systems in Relativity

Now that we have a qualitative understanding of what happens to length, time, and simultaneity as velocities approach the speed of light, it is useful to combine these so that we have equations by which we can convert from one observer's coordinates to another observer's coordinates. This will allow us to figure out where and when events seen by one observer will be seen by another observer. So if Sarah records something happening at a particular place and time, what place and time will Michael record for the event? Again, this refers to when the event actually happened, as computed by Sarah and Michael respectively – not merely when they see the light arrive.

In our analogy with the train, we took Sarah to be stationary relative to Earth's surface. We will call this the O reference frame. She measures the position of any event with coordinates x, y, z, and measures the time as t. By convention, Sarah is always located at the origin of her reference frame $x = 0$, $y = 0$, $z = 0$. An event is anything that happens anywhere in space, such as a lightning strike. Every event happens at a position described by the three spatial coordinates: x, y, and z. Every event also happens at a particular time, which we describe by the time coordinate t. Thus, every event that happens has four coordinates (x, y, z, t).

Michael's train moves in the positive-x direction at speed v. His reference frame is O'. He measures the position of an event with coordinates x', y', z', and measures the time as t'. By convention, his position is always at the origin of O', with coordinates $x' = 0$, $y' = 0$, $z' = 0$.

The four coordinates that mark any event as measured by Michael are x', y', z', t'.

At the moment Michael passes by Sarah, both observers agree to mark this as time zero. So $t = t' = 0$. We'll neglect the small distance between Sarah and the train track, such that Michael and Sarah have essentially the same position at $t = t' = 0$. For that instant, both observers are at $x = 0$, $y = 0$, $z = 0$, $t = 0$, and $x' = 0$, $y' = 0$, $z' = 0$, $t' = 0$. Our goal is to derive equations that allow us to convert from Sarah's coordinates for any event (x, y, z, t) to Michael's coordinates (x', y', z', t') or the reverse.

The Galilean Transformation

If the speed of Michael's train is very small, then relativistic effects like time dilation and length contraction are negligible. Under such circumstances, the time Michael records for any event will be about the same as what Sarah records. So $t = t'$. And since Michael moves only in the x-direction, there will be no difference in the y or z directions between Michael and Sarah. So $y = y'$ and $z = z'$. The only difference between x and x' will be due to the increasing distance between Michael and Sarah, which is vt. Therefore, $x' = x-vt$. These four equations together form the Galilean transformation:

$$x' = x - vt \qquad (8.1)$$
$$y' = y$$
$$z' = z$$
$$t' = t$$

These equations work very well as an approximation when the relative velocity between the two reference frames is much less than the speed of light. But at higher speeds, length contraction, time dilation, and the relativity of simultaneity become significant. So the Galilean transformation becomes increasingly inaccurate as velocity becomes a substantial fraction of the speed of light.

Our goal is to discover the true relationship between Michael's coordinates (O') and Sarah's (O) that is accurate for all speeds. This relationship is called the *Lorentz transformation*. To discover this, we will use the two postulates of relativity, along with the time dilation and length contraction formulas that we derived in earlier chapters. Fortunately, the relationship between y and y' coordinates is easy, and so is the relationship between z and z'. Recall from the "painting the hallway" thought experiment that motion along the x-direction does not affect measurement of widths or heights in the y and z directions respectively. Therefore, these will be unchanged from the Galilean transformation: $y' = y$, and $z' = z$. But since lengths and times are affected, the derivation of x' and t' will be more involved. There are several different ways to derive the Lorentz transformations. Below is a method that makes use of our previous discoveries involving length contraction and time dilation.

Derivation of the Lorentz Transformation

As before, Michael is at the middle of his train at $x' = 0$. From Sarah's perspective, he and his train are moving at speed v. So his position at any time, as measured by Sarah, is $x_M = vt$. At some positive time (t), suppose lightning strikes the front of Michael's train. From his perspective, this happens at time t', and position x', where x' is the distance from Michael to the front of his train. From Sarah's perspective the lightning strike occurs at time t and position x.

From Michael's perspective, the lightning strike is a distance of x' away from him. But from Sarah's perspective, Michael's train is contracted by the Lorentz factor γ. Recall that the Lorentz factor is:

$$\gamma = \frac{1}{\sqrt{1 - \dfrac{v^2}{c^2}}}$$

So from her perspective, the distance of the lightning strike from Michael will be x'/γ. Adding this distance to Michael's distance from Sarah, we find that the position of the lightning strike relative to Sarah will be $x = x_M + x'/\gamma$. And since $x_M = vt$, we find:

$$x = vt + x'/\gamma$$

Solving this for x', we find:

$$x' = \gamma(x - vt)$$

We now have the relationship between x' and x. What remains is to find the relationship between t' and t. To accomplish this, we consider the time (according to Sarah) t_A when the light from the lightning bolt arrives at Michael's position, and compare this to Michael's recorded time of arrival t_A'. This is conceptually simple, but requires a bit of algebra.

From Sarah's perspective, the distance between Michael and the lightning bolt is $x-vt$, and the speed of the light beam plus the speed of Michael is $c+v$. So from her perspective, the light will take a time of $(x-vt)/(c+v)$ to reach Michael. And since the lightning bolt struck at time t, from Sarah's perspective the light reaches Michael at time:

$$t_A = t + \frac{x - vt}{c + v}$$

Multiplying the second term by $c-v/c-v$ we obtain:

$$t_A = t + \frac{cx - vct}{c^2 - v^2} + \frac{v^2 t - vx}{c^2 - v^2}$$

When we divide both the numerator and denominator of the middle term by c^2, we see that the square of the Lorentz factor appears. And we also see the expression $x-vt$, which we can substitute from our formula above relating x' to x.

$$\frac{cx - cvt}{c^2 - v^2} = \frac{\gamma^2 (x - vt)}{c} = \frac{\gamma x'}{c}$$

So the arrival time according to Sarah's clock is:

$$t_A = t + \frac{\gamma x'}{c} + \frac{v^2 t - vx}{c^2 - v^2}$$

From Michael's reference frame, the light has to travel a distance of x' and travels at speed c. Adding this to the time the lightning bolt struck according to Michael's clock gives the arrival time:

$$t_A' = t' + \frac{x'}{c}$$

From Sarah's perspective, Michael's clock has been ticking slowly by the Lorentz factor. Therefore, the time he reads on his clock when the light arrives will be less than the time Sarah reads on hers by γ.

$$t'_A = t_A/\gamma$$

Substituting our previous equations for t_A' and t_A into this one, we find:

$$t' + \frac{x'}{c} = \frac{t}{\gamma} + \frac{x'}{c} + \frac{v^2 t - vx}{\gamma(c^2 - v^2)}$$

Subtracting x'/c from both sides, and combining the remaining terms gives:

$$t' = \frac{tc^2 - vx}{\gamma(c^2 - v^2)} = \frac{t - \frac{vx}{c^2}}{\gamma\left(1 - \frac{v^2}{c^2}\right)} = \frac{\gamma^2\left(t - \frac{vx}{c^2}\right)}{\gamma}$$

This finally simplifies to:

$$t' = \gamma\left(t - \frac{vx}{c^2}\right)$$

This formula relates t' and t in terms of known variables v, x, and γ.

Together with the previous formulas for the y and z axes, we have the complete Lorentz transformation:

$$
\boxed{
\begin{aligned}
x' &= \gamma(x - vt) \\
y' &= y \\
z' &= z \\
t' &= \gamma\left(t - \frac{vx}{c^2}\right)
\end{aligned}
}
\qquad (8.2)
$$

With a bit of algebra, it is possible to obtain the *inverse* Lorentz transformation from these, which gives Sarah's coordinates (O) in terms of Michael's (O').

$$x = \gamma(x' + vt') \tag{8.3}$$
$$y = y'$$
$$z = z'$$
$$t = \gamma\left(t' + \frac{vx'}{c^2}\right)$$

Notice that the Lorentz transformation is nearly identical to its inverse transformation. This would have to be the case due to the first postulate of relativity: the laws of physics work the same for Michael as they do for Sarah. The only difference is that from Michael's perspective, Sarah is moving at speed v in the *negative* x-direction. Thus, there is a sign difference in the term with the velocity v. The equations are otherwise identical in form.

Now that we have the Lorentz transformation, we can convert the coordinates from any reference frame to any other reference frame. The above four equations are sometimes also called a Lorentz boost, because they are a special case of the Lorentz transformation where the spatial coordinates are not rotated relative to each other.

Constraining motion to the x-direction does not limit application in any way, because we can always choose our axis directions. If we are concerned about motion in a different direction, we can use trigonometry to rotate the coordinate system so that the new x-axis is the direction of motion. (The details are provided below.) We then apply the Lorentz boost. If necessary, we can then rotate back to the original coordinate system.

The Lorentz transformation includes all the relativistic effects we have so far encountered. It takes into consideration length contraction, time dilation, and the relativity of simultaneity. Using the Lorentz transformation, we could revisit the "trapping the train" paradox from the previous chapter. Recall, we found that if the doors close at the same time from Sarah's reference frame, then they are not simultaneous from Michael's reference frame. The Lorentz transformation will allow us to compute exactly when each door closes from Michael's perspective, if we are given Sarah's perspective (or vice versa).

An interesting historical note is that the Lorentz transformation was derived *before* Einstein discovered relativity. The Dutch physicist Hendrik Lorentz derived these equations to explain how Maxwell's equations could appear to work identically in all inertial reference frames. It seems that Lorentz himself did not fully understand the significance of these equations. He was still under the impression that light required a luminiferous ether in which to travel, and that this ether represented an absolute reference frame. Einstein later showed that these equations can be derived from the postulates of relativity, just as we have done in this chapter.

Relativistic Velocity Addition

Previously, we discussed the Galilean velocity addition theorem. This is used when we have an observer moving at speed v relative to Earth, and another object moving at speed w faster than the observer. According to the Galilean velocity addition theorem (equation 2.1), the speed of the object relative to Earth will be:

$$W = v + w$$

This formula works perfectly well when both velocities are measured by the same observer. That is, if Sarah measures Michael's speed as v, and she measures the speed of a baseball as w faster than Michael, then indeed the ball's speed relative to Sarah will be $v+w$. However, this formula fails when the two velocities have been measured by different observers, though it still works as an approximation when all velocities are very small relative to the speed of light. That is, we may not add *Michael's* estimation of the baseball's speed to *Sarah's* estimation of Michael's speed, to find Sarah's estimation of the ball's speed – at least not when the velocities are a substantial fraction of the speed of light.

The reason that we may not simply add velocities that have been measured in two different reference frames is because the two different observers have a different standard of length, time, and simultaneity, due to the effects of length contraction and time dilation. Using the Lorentz transformation, we can now derive the correct relativistic velocity addition theorem that will allow us to compute the speed of the baseball relative to Sarah, if we know the speed of the ball relative to Michael (w) and Michael's speed relative to Sarah (v).

The speed of the baseball relative to Michael will be the distance it travels ($\Delta x' = x_1'-x_0'$) divided by the time it takes to travel that distance ($\Delta t' = t_1'-t_0'$). So we have:

$$w = \frac{\Delta x'}{\Delta t'}$$

From Sarah's perspective, the speed of the baseball will be the distance it travels ($\Delta x = x_1-x_0$) divided by the time it takes to traverse that distance ($\Delta t = t_1-t_0$). So we have:

$$W = \frac{\Delta x}{\Delta t} = \frac{x_1 - x_0}{t_1 - t_0}$$

From the inverse Lorentz equations (8.3), we substitute the values of x_1, x_0, t_1, and t_0 to get these in terms of the primed coordinates, and the rest is algebra:

$$W = \frac{\gamma(x_1' + vt_1') - \gamma(x_0' + vt_0')}{\gamma\left(t_1' + \frac{vx_1'}{c^2}\right) - \gamma\left(t_0' + \frac{vx_0'}{c^2}\right)}$$

The Lorentz factor (γ) appears in both the numerator and denominator, so it cancels. And rearranging terms yields:

$$W = \frac{x_1' - x_0' + v(t_1' - t_0')}{t_1' - t_0' + \frac{v}{c^2}(x_1' - x_0')} = \frac{\Delta x' + v\Delta t'}{\Delta t' + \frac{v}{c^2}\Delta x'}$$

Dividing numerator and denominator by $\Delta t'$ gives the following:

$$W = \frac{\frac{\Delta x'}{\Delta t'} + v}{1 + \frac{v}{c^2}\frac{\Delta x'}{\Delta t'}}$$

And since $w = \Delta x'/\Delta t'$, by substitution we arrive at the answer:

$$W = \frac{v + w}{1 + \frac{vw}{c^2}} \qquad (8.4)$$

With the relativistic velocity addition formula, we can now directly add velocities from two different reference frames, without having to explicitly compute any time dilation or length contraction effects. These are built into the formula.

So if Michael's train travels at half the speed of light relative to Sarah, and he throws a baseball forward at half the speed of light relative to himself, what is the speed of the ball relative to Sarah?

Here, $v = w = \frac{1}{2}c$. So $W = (\frac{1}{2}c + \frac{1}{2}c)/(1 + \frac{1}{2}c\frac{1}{2}c/c^2) = (4/5)c$. Sarah will see the ball travel at 80% the speed of light.

Notice that the formula also works if one (or both) of the velocities is actually the speed of light. If Michael travels at half the speed of light relative to Sarah ($v = \frac{1}{2}c$), and shines a beam of light forward that travels c faster than him (so $w = c$), then what is the speed of the beam relative to Sarah? Putting these values into the formula yields: $W = (\frac{1}{2}c + c) / (1 + \frac{1}{2}c^2/c^2) = c$. The beam will travel at exactly the speed of light relative to Sarah. This has to be the case because of the second postulate of relativity. But this is a nice confirmation that the relativistic velocity addition formula is correct.

Implications of the Lorentz Transformation

When some event happens at location x, y, z and at time t from Sarah's reference frame, we can compute what the location x', y', z' and time t' will be from Michael's reference frame for the same event. But when we look at the formula for x' and t', we notice something extraordinary. The formula for x' depends not only on x but also on t – Sarah's time. Of course, this was also true in the Galilean transformation and perhaps merely reflects the fact that the distance between Michael and Sarah increases with time. But, remarkably, the formula for t' depends not only on t but also on x. The implication is profound: what Michael sees as time is what Sarah sees as a combination of time and space, and what Michael sees as space is what Sarah sees as a combination of space and time.

It is as if space and time are not the separate things they seem to be. Rather, what we perceive to be space or time depends on our velocity. There is another situation where the coordinates assigned by one person can be mixed, such that what one person sees as entirely x', the other person sees as a combination of x and y. This situation is *rotation*. Consider the following illustration.

Joe and Stacy are two observers at the same location and both are stationary. Joe defines his coordinate system such that the y-axis points directly north. Stacy, however, defines her coordinate system such that the y'-axis points toward her hometown, which is to the north but slightly west of her position. So her coordinate system is rotated with respect to Joe's coordinate system by angle θ as shown in the figure.

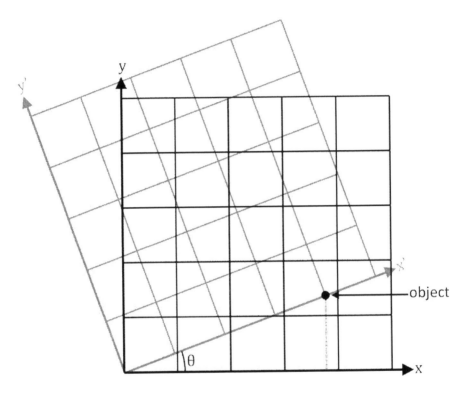

Now consider some object which Stacy measures to be at position x' = 4, y' = 0. In her coordinate system, the object is entirely on the x'-axis, with a value of zero for the y' coordinate. But in Joe's coordinate system, the position for the object has positive value for both x and y. In fact, his measurement for the x-coordinate will be somewhat less than Stacy's value for x', and his y value will be greater than Stacy's y'. So what was entirely x' for Stacy is a combination of x and y for Joe.

For any rotation, the x'-coordinate in the rotated system will be determined by a combination of x and y in the non-rotated system:

$$x' = Ax + By$$

The constants A and B will depend on the angle (θ). Likewise, the y'-coordinate in the rotated system will be determined by a combination of x and y in the non-rotated system, where C and D depend on the angle:

$$y' = Cy + Dx$$

For simple rotations like the kind illustrated in the figure, the values of the constants are found by trigonometry to be: $A = C = \cos(\theta)$, $B = -\sin(\theta)$, $D = \sin(\theta)$. So we have:

$$x' = x\cos(\theta) + y\sin(\theta) \qquad (8.5)$$
$$y' = y\cos(\theta) - x\sin(\theta)$$

From Stacy's coordinate system, this particular object's position is determined entirely by x'. But from Joe's coordinate system, its position is a combination of x and y. This is analogous to what we found when we compared Sarah's stationary reference frame to Michael's moving frame. Much as the x-coordinate and y-coordinate become "mixed" when we shift to a rotated frame, so space (the x-coordinate) and time (the t-coordinate) become "mixed" when we shift to a moving frame. *Velocity is much like a rotation between space and time.*

Joe and Stacy will disagree on the coordinates of the object because they are using different coordinate systems. But notice that they will agree on the distance to the object. The distance formula computes distance (r) in terms of lengths x and y (or x' and y') and follows from the Pythagorean theorem:

$$r = \sqrt{x^2 + y^2}$$

Clearly, the object in the above figure is exactly four units away from both Joe and Stacy. While they disagree on the relative contributions of x and y, they agree on a particular *combination* of those coordinates – the total distance. This prompts us to ask if there might be something similar in relativity. Though Michael and Sarah disagree on the time and position of a particular event, is there some combination of time and space upon which they would agree? If so, then space and time are somewhat subjective aspects of an objective reality that encompasses both. This will be the subject of the next chapter.

Chapter 9

Spacetime

In the previous chapter we found that what a stationary observer perceives as space only (x-coordinate), a moving observer will perceive as a combination of space (x'-coordinate) and time (t'-coordinate). This suggests that space and time are not independent structures, but are two parts of a larger whole: spacetime. Which aspect of spacetime we see depends on our state of motion.

This is analogous to comparing the x and y coordinates of one system to the x' and y' coordinates of a rotated system. Yet, the distance between any two objects does not change when we rotate the coordinate system. Though the coordinates (x, y) may differ (from x' and y'), the underlying reality – distance from the origin – does not change. There is something similar involving space and time. But let's first review how distances and coordinates work under rotation.

Consider two different points with locations (x_0, y_0, z_0) and (x_1, y_1, z_1) respectively. When comparing two different values of the same type, such as the x-coordinate of two different points, we often use a convenient shorthand notation. We express the difference between two values with an uppercase delta symbol as follows:

$$\Delta x = x_1 - x_0 \qquad (9.1)$$
$$\Delta y = y_1 - y_0$$
$$\Delta z = z_1 - z_0$$

We can also do this for two different points in time:

$$\Delta t = t_1 - t_0 \tag{9.2}$$

Using this notation, for any two points in space, the distance between them (r) is expressed as follows:

$$r = \sqrt{\Delta x^2 + \Delta y^2 + \Delta z^2} \tag{9.3}$$

This is simply the more general three-dimensional version of the distance formula covered in the last chapter, but for any two points in space.   The formula works whether we use primed or unprimed coordinates. The figure below illustrates how the distance between any two points must be the same whether we use the non-rotated coordinate system or the rotated coordinate system.

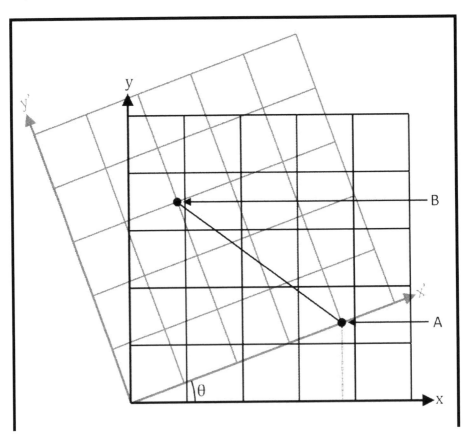

Consider the positions of object A and object B in the figure. In the primed (rotated) coordinate system, A is located at $x_0' = 4$, $y_0' = 0$, and B is located at $x_1' = 2$, $y_1' = 3$. We will take all z' positions to be zero for this illustration. So we have $\Delta x' = 2$, $\Delta y' = 3$, $\Delta z' = 0$. The angle of rotation (θ) is 20°. Using the distance formula (9.3), we find the distance between these two points to be:

$$r = \sqrt{2^2 + 3^2} = \sqrt{13} \approx 3.61$$

If we use the non-rotated system our values for Δx and Δy will differ. But we will compute the same distance between these two points. The angle of rotation between these two coordinate systems is 20 degrees. From the rotation equations (8.5) covered in the previous chapter, we find that $x_0 \approx 3.759$, $y_0 \approx 1.368$, and $x_1 \approx 0.853$, $y_1 \approx 3.503$. So we have $\Delta x \approx -2.906$, $\Delta y \approx 2.135$. Substituting these values into the distance formula (9.3) we find

$$r \approx \sqrt{(-2.906)^2 + 2.135^2} \approx 3.61$$

We can see that distances are unchanged under rotation.

The Spacetime Interval

This brings us to the concept of the *spacetime interval*. Just as the distance between two points is unchanged by coordinate rotation (even though the specific coordinates of each point differ), the spacetime interval is unchanged by a Lorentz transformation, even though the spacetime coordinates will differ between the two reference frames. The spacetime interval is defined as follows:

$$s = \sqrt{\Delta x^2 + \Delta y^2 + \Delta z^2 - c^2 \Delta t^2} \qquad (9.4)$$

Notice that this is eerily similar to the distance formula, but with an extra term for the time coordinate. We may therefore think of the spacetime interval as a sort of "distance" through spacetime rather than just an ordinary distance through space. In fact, when two events are simultaneous, (in which case $\Delta t = 0$), the spacetime interval is identical with the ordinary distance between the two events.

It may look as though the time difference (Δt) is treated differently, since it is multiplied by the speed of light squared but the other coordinates are not. However, this is merely a unit conversion. The time coordinate can only be combined with the spatial coordinates if all are in the same units, such as meters. But we normally measure time in seconds, not meters. Multiplying by the speed of light squared converts the time coordinate to units of length squared, so that these quantities can be added together naturally. If we used the same type of units to measure distance as we use to measure time, we would not need to have this conversion factor.

However, one difference concerning the time coordinate cannot be ignored: the sign change. The spatial coordinates squared contribute in a positive way to the spacetime interval, whereas the time coordinate squared *is subtracted* from the others. This is very significant. Time is a dimension, and is just like length, width, and height in most respects. But it is somehow "opposite" from the three spatial dimensions. An increase in Δx, Δy, or Δz will increase the spacetime interval. But an increase in Δt will *decrease* the spacetime interval.

Our universe is a four-dimensional structure; it takes four coordinates (x, y, z, t) to describe the position and time of any event in spacetime. But the fourth dimension (t) is somehow different from the other three. That little negative sign in the spacetime interval equation is why we experience time differently than we experience space.

The spacetime interval is invariant under the Lorentz transformation. This means that whatever one inertial observer calculates the spacetime interval to be between any two events, all other inertial observers will calculate exactly the same value. You can verify this for yourself using the Lorentz transformations from the previous chapter. Pick two points in spacetime with any values you like (x_0, y_0, z_0, t_0) and (x_1, y_1, z_1, t_1). Then calculate the spacetime interval (s). Now calculate the position of the same two points in the O' frame selecting a velocity of your choice (v) using the Lorentz transformation to find (x_0', y_0', z_0', t_0') and (x_1', y_1', z_1', t_1'). If you again compute the spacetime interval (s) using the primed coordinates, you will find it is always exactly the same as the value you computed from the unprimed coordinates.

So Michael and Sarah may disagree on the times and positions of any events in the universe. But they will always agree on the spacetime interval between any two events. This tells us something quite profound about the universe. The spacetime interval is the "real" absolute quantity. The relative contributions of space and time to the spacetime interval are observer-dependent. What we see as space or as time depends on our velocity. But the spacetime interval itself is absolute and totally independent of the observer's state of motion.

This refutes a common misconception about the theory of relativity – the notion that relativity denies absolute quantities and reduces to philosophical relativism. That simply isn't so. Relativity does have absolutes, such as the speed of light in vacuum and the spacetime interval. These are invariant absolute quantities. It simply challenges our intuitive expectation about *which* things are relative to the observer and which things are absolute and invariant.

The Range of the Spacetime Interval

By convention, the spacetime interval is always taken to be the principle (positive) square root of the variables under the radical. So it is never negative. It can be zero. And it can also be "imaginary": that is, it can be *the square root of a negative number*. This requires some explanation.

The so-called "imaginary numbers" are neither positive, nor negative, nor zero. You can think of them as being along a perpendicular axis to the number line, passing through zero. But the square of an imaginary number is negative. So they can be hard to visualize, but they do exist. The term "imaginary" is therefore very misleading. To add insult to injury, numbers whose square is positive are called "real." But these are just names. The existence of the "imaginary" numbers is just as legitimate as the existence of the "real" numbers.

The spacetime interval can be positive, zero, or imaginary. For example, suppose $\Delta x^2 + \Delta y^2 + \Delta z^2 - c^2 \Delta t^2$ = -4. In that case, $s = \sqrt{-4}$. The solution is $s = 2i$, where i is defined to be $i = \sqrt{-1}$. Since some people prefer to stick with the so-called "real" numbers, it is often customary to define the spacetime interval by its square, as follows:

$$s^2 = \Delta x^2 + \Delta y^2 + \Delta z^2 - c^2 \Delta t^2 \qquad (9.5)$$

This ensures that all resulting values are "real." The square of the spacetime interval can be positive, negative, or zero. The spacetime interval can be thought of as a distance in space minus the "distance" in time between any two events. The important thing to remember about the spacetime interval is that the time term has the opposite sign as the three spatial terms. There is no set convention on whether the spatial terms should be positive and the time term negative (as I have listed above), or the reverse. That is, about half of physicists define the spacetime interval as:

$$s^2 = -\Delta x^2 - \Delta y^2 - \Delta z^2 + c^2 \Delta t^2 \qquad (9.6)$$

Either convention is perfectly acceptable. Just be consistent. I prefer to have the spatial terms positive so that when the time difference is zero the spacetime interval is equal to ordinary distance.

If two events are widely separated in space, but happen at nearly the same time, then they are called "spacelike" events. Under the convention we use in this text, s^2 will be positive for these. For spacelike events,

$$\Delta x^2 + \Delta y^2 + \Delta z^2 > c^2 \Delta t^2$$

Conversely, if two events happen at nearby locations but at very different times, then they are called "timelike" events. Under our convention, s^2 will be negative for these events. For timelike events,

$$\Delta x^2 + \Delta y^2 + \Delta z^2 < c^2 \Delta t^2$$

If two events are exactly as far in space as they are in time from each other, then the spacetime interval will be zero. This is the case for events where a light pulse emitted at one event can just reach the other event as it happens. These are called "lightlike" events. For lightlike events,

$$\Delta x^2 + \Delta y^2 + \Delta z^2 = c^2 \Delta t^2$$

Since the spacetime interval is invariant under the Lorentz transformation, when one observer computes that two events are spacelike, all other observers will also find those same two events to be spacelike. Likewise, what is timelike for one observer will be timelike for all other observers. And what is lightlike for one observer will be lightlike for all. The spacetime interval therefore divides all of spacetime into three regions. This is illustrated visually below.

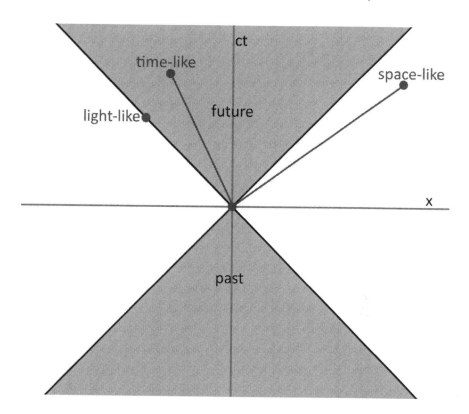

The figure depicts two dimensions of spacetime, the x-dimension and the (time) t-dimension. This representation of spacetime is called a Minkowski diagram, after Hermann Minkowski who first developed it in 1908. By convention, the time dimension is always shown on the vertical axis, with past events below the origin and future events above. The observer is at the origin at $x = 0$, and the present time is $t = 0$. Note that the time axis is multiplied by c so that its units are the same as the x-axis. When the units are the same, light always travels at a 45° angle. The bottom two 45° line segments represent two beams of light that arrive at the observer's position at time = 0. These form the "past light cone" because when the y-dimension is included, the shape is a complete cone rather than just two lines. The top two 45° line segments represent light beams that are emitted by the observer at time $t = 0$. These form the "future light cone."

All points in the shaded regions (within the light cones) are timelike with respect to the origin. There is a greater "distance" in time between these points and the origin than the distance in space. So $s^2 < 0$. All points in the unshaded regions are spacelike with respect to the origin. Their distance in space from the origin is greater than their "distance" in time. So $s^2 > 0$. Points that are exactly on either light cone are lightlike with respect to the origin. So $s^2 = 0$ for these, as is always the case for light.

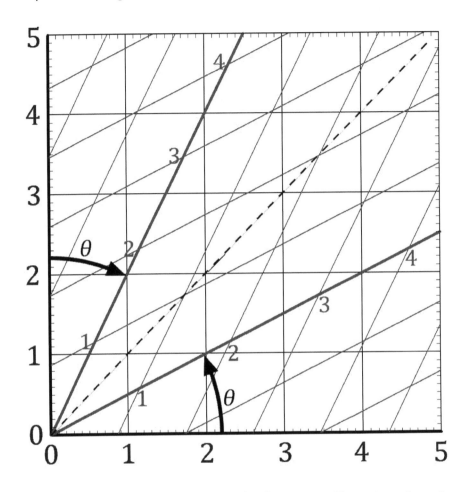

We can also use a Minkowski diagram to illustrate a Lorentz transformation as shown in the above figure. The x and y axes are shown in black, with the x' and y' shown in light grey for an object moving in the positive x direction at 50% the speed of light. We can see that velocity acts very much like a rotation in space and time with the exception that the time axis rotates in the opposite direction from the space axis! That is, in a standard rotation, if the x' axis is rotated counterclockwise relative to the x axis, then so is the y' axis relative to the y axis. But here we see that the t' axis is rotated *clockwise* relative to t. The other difference is that the x' and t' coordinates are stretched relative to x and t. A light beam emitted at $x = t = 0$ is shown by the dashed line. The angle of rotation (θ) is the inverse tangent of v/c, which is approximately 26.57° for an object moving at $0.5c$.

We can see that this figure illustrates a Lorentz transformation by selecting a point and running the numbers. For example, consider an event at $x = 4$, $t = 2$. This point lies directly on the x' axis, so the t' coordinate will be zero. And the x' coordinate appears to be a bit less than 3.5. Using the Lorentz transformation, we find that indeed $x' \approx 3.4641$. As we consider objects of higher speed, the angle (θ) increases. But the spacetime interval between any two events is always the same in either reference frame.

Proper Time

Before moving on to the application of the spacetime interval, there is an important related concept we must discuss: *proper time.* Designated by the lowercase Greek letter tau (τ), proper time is the time between two events that would be measured by an inertial clock that is at the location of both events when they happen. If you were to run a race (in a straight line at constant speed), and time yourself, the proper time between when you started and when you finished would be the time you recorded on your stopwatch. This might be slightly different from the time recorded by bystanders because of the effects of time dilation and the relativity of simultaneity. But since your clock was present at the location of the starting line (event A) and at the finish line (event B), your clock records the "proper time."

What is the relationship between proper time and the spacetime interval? Really, they are the same thing. Both represent the "distance" through spacetime between two events, with the time term having the opposite sign as the space terms. They simply express this "distance" in different units. The spacetime interval is given in units of *distance*, such as meters or miles. Recall, we had to convert the time term to units of distance in the spacetime formula by multiplying by c^2 so that we could add it to the spatial terms.

The proper time expresses the same "distance" through spacetime but with units of time, like seconds. Therefore, all the distance terms must be divided by c^2 so that they are expressed in units of time. Like the spacetime interval, proper time is often given by its square so that imaginary numbers are not needed. The proper time between two events is defined by this formula:

$$\tau^2 = \Delta t^2 - \frac{\Delta x^2}{c^2} - \frac{\Delta y^2}{c^2} - \frac{\Delta z^2}{c^2} \qquad (9.7)$$

Generally, when expressing proper time, the time term is positive and the spatial terms are negative. Recall that there is no consensus convention on the sign of the terms for the spacetime interval; the only requirement is that the time term has the opposite sign as the three spatial terms. Under the convention I prefer, the spatial terms are positive for s. Therefore, the square of proper time will have the *opposite* sign as the square of the spacetime interval:

$$\tau^2 = -\frac{s^2}{c^2} \qquad (9.8)$$

But using the other convention, the proper time and spacetime interval have the same sign:

$$\tau^2 = \frac{s^2}{c^2}$$

Order of Events

Light cones are very important in the physics of relativity, particularly when it comes to cause-and-effect. They are the boundary between the spacelike regions and timelike regions with respect to a given reference point. When two events happen, it is important to know whether they are spacelike events or timelike events (or lightlike events) if we are concerned about causality or the order in which these events happen. The order in which two timelike events occurs is invariant under a Lorentz transformation. But the order of two spacelike events is not. What does this mean?

Suppose an observer sees two events. He finds that event A happened first, and event B happened at a later time. That doesn't necessarily mean that he *sees* A first. Rather it means that after subtracting off the light travel time from the source, he computes that event A actually happened before event B. And suppose after measuring distances and times, he finds that A and B are timelike with respect to each other. That means their "distance" in time is greater than their distance in space. Under these circumstances, all other inertial observers in the universe, whatever their velocity, will agree that event A happened before event B.[11] They might calculate different times and distances due to time dilation and length contraction. But all will agree on the order in which the events actually occurred.

Now consider a different scenario. The observer sees two events and computes that A occurred before B. But this time, suppose that these two events are spacelike relative to each other. In that case, not all inertial observers will agree that A happened before B. Some will agree. Others will compute that B actually happened before A. And some observers will claim that A and B are simultaneous. The order of spacelike events is observer-dependent. In fact, *for every two spacelike events, there is a reference frame in which the two events are simultaneous.*

We saw an example of this in our discussion of the "trapping the train" thought experiment. We found that while Sarah computes that the two doors on the tunnel closed at the same time, Michael found that B closed before A. Furthermore, a train traveling in the opposite direction would compute that door A closed before door B.

It is easy to show that these two doors represent spacelike events when we consider Sarah's perspective. From her point of view, they close simultaneously. So $\Delta t = 0$. Therefore, the distance between the doors in space will necessarily be greater than their "distance" in time, since the latter is zero.

We can use the Lorentz transformation to prove that timelike events always have the same order for any inertial observer, provided that such an observer moves slower than the speed of light: $v < c$.

Timelike events, by definition, have a greater "distance" in time than in space. Recall:

$$c^2 \Delta t^2 > \Delta x^2 + \Delta y^2 + \Delta z^2$$

For convenience, let's consider two events separated only by their x-coordinates. So $\Delta y = 0$ and $\Delta z = 0$. Then we find:

$$c^2 \Delta t^2 > \Delta x^2$$

---

[11] This presupposes that the observers are moving slower than the speed of light. We will examine the possibility of faster-than-light travel in a later chapter.

And let's define our origin so that it coincides with one of the two events for both observers. So for event A, we have $t = t' = 0$, $x = x' = 0$. Then, the difference between the coordinates of the two events will simply be the coordinates of event B. So we have $\Delta x = x$, and $\Delta t = t$. Substituting into the previous formula, we find:

$$c^2 t^2 > x^2$$
$$t^2 > \frac{x^2}{c^2}$$

Taking the principle square root of both sides, we find:

$$|t| > \left|\frac{x}{c}\right|$$

The absolute values are necessary to preserve the inequality, because t could be either positive or negative, and so could x. All we know is that the square of t is greater than the square of x/c. Thus, t has a greater absolute magnitude than x/c, though it might have the opposite sign. Since the left-hand side of the inequality is greater than the right-hand side, we can multiply the right-hand side by any quantity less than one, and the inequality will still be true. And since $v<c$. the quantity v/c must be less than one. Therefore, the inequality is still true if we multiply the right-hand side by v/c:

$$|t| > \left|\frac{vx}{c^2}\right|$$

Recall from the Lorentz transformation:

$$t' = \gamma\left(t - \frac{vx}{c^2}\right)$$

But from our inequality, we know that the absolute magnitude of the left term in the parenthesis (t) must always be greater than the absolute magnitude of the right-hand term (vx/c^2). That means the sign of the quantity in parenthesis will be determined *entirely* by t. For example, if t were 5, then (vx/c^2) would have to be something less than 5. So 5 minus something less than 5 is still positive. Conversely, if t were -10, then (vx/c^2) would have an absolute magnitude less than 10. So even if vx/c^2 were negative, and thus subtracting a negative adds to the sum, the quantity ($t - vx/c^2$) would still have to be negative. Moreover, the Lorentz factor (γ) is always positive, being some number greater than one. So it does not affect the sign of the right-hand side.

Therefore, t' must always have the same sign as t for timelike events. If event A happens before event B, then t is positive; therefore, t' must also be positive. If event B happens before event A, then t is negative; therefore, t' must also be negative. Both the stationary and moving observer will agree on the order in which any two timelike events occur.

A similar proof can establish the reverse for spacelike events. Simply flip the inequality above so that the space term is a larger magnitude than the time term. The result will show that the right-hand term in the parenthesis (vx/c^2) will be larger in absolute magnitude than t, and will therefore dominate the sign of the right-hand side. Thus, t' and t do not necessarily have the same sign for spacelike events. One could be positive while the other is negative.

The above proof assumes that velocity v is less than c. All observers for whom that is true will agree on the order of timelike events. But if v is allowed to be greater than c, then such an observer might disagree on the order of timelike events. That is, a person traveling faster than light would not always see timelike events happen in the same order as observers traveling slower than light. This is a profound result because timelike events may be *causal*.

Causal events are those where one event is the cause of another. We all have some experience with cause and effect. The effect follows in time from the cause, and does so in a way that is necessary – not just coincidental. If you drop a plate, it falls to the floor. The fall happens after you let go of the plate, and as a result of it. Letting go of the plate is one event. The plate falling to the floor is another event that follows in time. As far as we know, all causally connected events are always timelike or lightlike. They are never spacelike.

That means if you cause a plate to break because you let go of it and it falls to the floor, every other inertial observer in the universe with any velocity $v < c$ will agree that you let go of the plate *before* it hit the floor. The order of timelike events is universal. Effects never happen before their cause in any inertial reference frame that is slower than the speed of light.

But, if we allow the velocity of an observer to be greater than light ($v > c$) then the order of timelike events is no longer preserved. An observer moving faster than light might conclude that the plate hit the floor *before* you dropped it! The order of causally-connected events is the same for all observers moving slower than c. But observers traveling faster than light would see some causally-connected events happen in the opposite order. Their direction of travel through time can be the reverse of ours! The implication is very clear: *if you can travel faster than light, you can travel "backward" in time.*

Chapter 10

How to Build a Time Machine

We have already seen that it is possible – theoretically – to travel into the distant future. Just travel into space at a very high fraction of the speed of light, and turn around and come back. While such speeds are currently unattainable, from physics we know that time travel into the future is possible in principle. But any self-respecting time machine ought to be able to travel back into the past as well. It would not be much fun to travel into the future if you could never come back to the present to tell anyone about it. Is it possible to build a machine that can travel both forward and backward in time? This would constitute a true time machine.

As we saw in the previous chapter, the theory of relativity allows for the possibility of traveling backward in time *if* it is also possible to travel faster than the speed of light. That's the good news. The bad news is that there are some compelling reasons to think that faster-than-light travel is not possible, at least not for information-bearing systems. We have seen hints of that in previous chapters, and we will go into greater depth in later chapters. Nonetheless, *hypothetically* if faster-than-light travel were possible then time travel into the past would also be possible. This is provable from the Lorentz transformation, and gives us a much more profound understanding of the nature of space and time. Let's explore how this would work.

Suppose that in the year 2101 NASA has developed a new type of spaceship – the *Phoenix* – capable of faster-than-light travel. By whatever mechanism, when the engines engage, this "warp" ship instantly jumps to $1.5c$ relative to its previous state of motion. Imagine that the *Phoenix* is initially stationary relative to Earth. The engines engage, and the ship is now traveling at 1.5c relative to Earth, and coasts at constant speed headed toward the Alpha Centauri star system. This star system is approximately 4.3 light-years away. Therefore, it will take the *Phoenix* 2.87 years to reach the system according to Earth clocks.[12]

From Earth's perspective, the ship is traveling in the normal forward-time direction, because it arrives at its destination *after* it departed. But notice that the departure of the *Phoenix* from Earth and the arrival of the ship at Alpha Centauri are spacelike events. That is, $\Delta x^2 > c^2 \Delta t^2$. In this case, Δx = 4.3 light-years, Δt = 2.87 years, and c = 1 light-year per year. We saw in the previous chapter that when two events are spacelike, not all inertial observers will agree on the *order* in which the events took place. Therefore, in some reference frames, the arrival of the *Phoenix* at Alpha Centauri will occur *before* its departure from Earth. From these reference frames, the ship is traveling backward in time!

So simply by traveling faster than the speed of light, the direction through time will necessarily be reversed *in some frames of reference*. This strange phenomenon arises from the relativity of simultaneity we discussed in chapter 7. In this particular case, the Earth is not one of those frames that finds the ship traveling backward in time. But some other observer in deep space could. What reference frames would see the ship reach its destination before its departure? To compute this, we consult the Lorentz transformation (8.2).

---

[12] We are tempted at this point to use the time dilation formula (5.4) to see how much time the passengers would experience during their journey. This would be 2.87 years divided by the Lorentz factor. But substituting 1.5c into the formula (5.2) reveals that the Lorentz factor is the reciprocal of the square root of *negative* 1.25, which is an imaginary number. It is hard to understand what this means in practice. How can a clock tick an imaginary number of seconds?

Recall from the previous chapter that the spacetime interval is positive for spacelike events and imaginary for timelike events. This is due to the nature of the spacetime interval; the square of each spatial term is *added* whereas the time term is *subtracted* when computing s^2. Apparently, the time dimension of our universe is much like a dimension of space multiplied by the imaginary number *I* and *vice versa*. This suggests an answer to our question about time-dilation at superluminal velocities. Apparently, when traveling faster than light, the role of space (in the x-direction) and time *reverse*.

This is also illustrated in a Minkowski diagram. As we consider objects of increasing velocity, the angle (θ) increases. The x' axis rotates counterclockwise and the t' axis rotates clockwise. As the velocity passes c, the x' and t' axis cross! At superluminal speeds the x' (space) axis becomes timelike, and the t' (time) axis become spacelike.

From Earth's reference frame, the departure time is t_0, and the arrival time is t_1. The departure position is x_0 and the arrival position is x_1. In this example, we will take v to be the velocity of some other observer in space, traveling at less than the speed of light $v < c$. We will label the speed of the *Phoenix* as w, which in this case is $1.5c$. We also know that:

$$\Delta x = x_1 - x_0 = 4.3 \text{ light years}$$
$$\Delta t = t_1 - t_0 = 2.87 \text{ years}$$
$$w = \frac{\Delta x}{\Delta t} = 1.5c$$

From some other reference frame (O'), the departure time will be t_0' and the arrival time t_1'. If the arrival time happens before the departure time, then $\Delta t'$ will be negative, and this reference frame will perceive the ship traveling backward in time. So we need to solve for v in the case where $\Delta t' < 0$. From the Lorentz transformation (8.2) we derived previously:

$$t' = \gamma\left(t - \frac{vx}{c^2}\right)$$

The change in time $\Delta t' = t_1' - t_0'$. Applying the Lorentz transformation to this we find:

$$\Delta t' = \gamma\left(t_1 - \frac{vx_1}{c^2}\right) - \gamma\left(t_0 - \frac{vx_0}{c^2}\right)$$

Simplifying:

$$\Delta t' = \gamma\left(\Delta t - \frac{v}{c^2}\Delta x\right)$$

Now that we have the expression for $\Delta t'$ in terms of known quantities, we solve for the case where $\Delta t' < 0$ to find those reference frames which will perceive the warp ship as traveling backward in time.

$$\gamma\left(\Delta t - \frac{v}{c^2}\Delta x\right) < 0$$

Solving the inequality for v yields:

$$v > \frac{\Delta t}{\Delta x}c^2$$

The *Phoenix* travels a distance Δx in a time of Δt from Earth's reference frame. Thus, its speed relative to Earth is $w = \Delta x / \Delta t$. Substituting this into the inequality, we find:

$$v > \frac{c^2}{w}$$

Thus, any observer traveling at a velocity (relative to Earth) greater than c^2/w in the same direction as the *Phoenix*, will observe the ship arrive at its destination *before* it departs.

In our case, $w = 1.5c$. Hence, any observer traveling faster than $\frac{2}{3}c$ in the same direction as the *Phoenix*, will observe that ship traveling backward in time. The extreme strangeness of this phenomenon prompts me to point out once again that we are not discussing merely what the high-speed observer would *see* through his or her telescope. This is not merely an optical illusion due to differences in time lag at which the light from various events finally reaches some observer. Rather, from the high-speed observer's reference frame, the warp ship actually arrives at its destination before it departed – genuinely traveling backward in time.

But from the perspective of Earth, the *Phoenix* has not traveled backward in time since it reached its destination at some positive time (2.87 years) *after* its departure. So this type of scenario wouldn't help us if we wanted to prevent World War II, or witness the construction of the great Pyramids of Egypt. To travel into Earth's past would be truly tantalizing. Is our warp ship capable of doing this? And if so, what would be the required conditions?

Visiting Earth's Past

Suppose that in January of the year 2100, NASA launches a slower-than-light ship – the *Palomino* – toward Alpha Centauri. This sub-light ship travels at a constant speed of $0.99c$, and therefore its journey lasts approximately 4.34 years, as measured by Earth clocks. Unfortunately, less than a month into the mission, the *Palomino*'s engines are critically damaged. Even worse, the ship does not contain the necessary supplies to repair the engines. This will make it impossible for the ship to stop at its destination. Instead, the *Palomino* will simply fly past Alpha Centauri at $0.99c$, and continue to drift into space forever.

But all is not lost. Fortunately, the faster-than-light warp drive for the *Phoenix* is invented shortly after the departure of the *Palomino*. The *Phoenix* is then launched as a rescue mission, in June 2101 (1.48 years after the *Palomino* launch). Activating its warp engine, the *Phoenix* jumps to a speed of 1.5c, and reaches Alpha Centauri at the same time the *Palomino* does. Instead of stopping, the *Phoenix* merely slows to 0.99c and docks with the *Palomino*. The crew of the *Phoenix* transfers the necessary equipment to the *Palomino*. Both crews working together are able to repair the engines. The *Palomino* will be able to complete its mission.

After undocking with the *Palomino*, the crew of the *Phoenix* is ready to return to Earth. Both ships are still coasting away from Earth at 0.99c. Rather than slowing to a stop, the *Phoenix* simply rotates so that it points toward Earth. The crew then engages the engine, and the *Phoenix* jumps to 1.5c relative to the *Palomino*. From the *Palomino*'s reference frame, the *Phoenix* is traveling toward Earth at 1.5c, but the Earth is moving away at 0.99c from the *Palomino* (which has not yet activated its braking engines). From the *Palomino*'s reference frame, the Earth, Alpha Centauri, and the local universe are moving at 0.99c, and so they will experience length contraction. The stationary distance of 4.3 light-years between Earth and Alpha Centauri will be reduced to 0.607 light-years in the *Palomino*'s frame. Since the *Phoenix* travels 0.51c faster than Earth recedes, it will reach Earth in 1.19 years from the *Palomino*'s frame.

But from Earth's reference frame, when will the Phoenix arrive? We use the Lorentz transformation to find the astonishing answer:

From the *Palomino*'s perspective, the *Phoenix* arrives at Earth in 1.19 years. But the Earth, traveling at 0.99c, has moved during this time from its initial distance of 0.607 light-years to 1.78 light-years from the *Palomino*'s position. So in the *Palomino* frame we have:

$$\Delta t = 1.19 \text{ years}$$
$$\Delta x = 1.78 \text{ light years}$$
$$v = 0.99c$$

Following the same steps as in the previous example, we again find from the Lorentz transformation (8.2):

$$\Delta t' = \gamma \left(\Delta t - \frac{v}{c^2} \Delta x \right)$$

Here, $\Delta t'$ represents the length of time it took for the *Phoenix* to travel from Alpha Centauri back to Earth. At $v = 0.99c$, we find that the Lorentz factor is: $\gamma=7.09$, and substituting in the values for Δt and Δx we find:

$$\Delta t' = -4.09 \text{ years}$$

The truly amazing thing about our result is that the travel time of the journey is *negative*. Recall that $\Delta t'$ is the arrival time minus the departure time, as measured in the moving frame (which in this case is Earth). A negative travel time therefore means that the arrival time is earlier than the departure time. As measured by clocks on Earth, the *Phoenix* arrives at Earth 4.09 years *before* it departed the Alpha Centauri system! The *Phoenix* has traveled backward in time.

The other extraordinary observation is the magnitude of the effect. Not only has the *Phoenix* traveled backward in time, it has traveled *four years* backward in time, according to clocks on Earth. This is extremely significant, because recall that the *Phoenix* took only 2.87 years to travel from Earth to Alpha Centauri, by Earth clocks. Neglecting the small amount of time to repair the *Palomino*, the total time for the *Phoenix* to travel to Alpha Centauri and back to Earth is: 2.87 + (-4.09) = -1.22 years: a negative number. That means, the *Phoenix* arrived back at an *earlier* time than its initial departure!

Following the mission order-of-events: the *Phoenix* departs in June, 2101. It arrives at Alpha Centauri in May 2104. But it arrives back on Earth in March 2100: over a year before its own launch date! This means that between March 2100 and June 2101, there would be *two Phoenix* spaceships on Earth, and *two* of each crew member. One of these ships and crews has already completed the mission, while the other crew is still training for it. And then the situation becomes even stranger.

The Temporal Paradox

When the *Phoenix* returns from its journey, in March 2100, the ship captain reports that the mission was a complete success. The *Palomino*'s engines have been repaired, and its mission to explore Alpha Centauri is on track.

The leaders at NASA are very pleased with this. So they decide that there is no need to launch the *Phoenix* rescue mission in June 2101, since the *Palomino*'s engines have been repaired. Why not instead use the *Phoenix* to visit a different star system, such as Barnard's star, since the *Palomino* is more than capable of mapping the Alpha Centauri system. This will give the crew further time to train, since there is no longer any urgency.

As a bonus, NASA is pleased that there are now *two Phoenix* starships and *two* crews: one that has already completed the *Palomino* rescue mission (let's call this *Phoenix B*), and the other that was preparing for it but will now instead visit Barnard's star (*Phoenix A*). So NASA decides to send the duplicate *Phoenix* to yet a third star system: Sirius. There will be no rescue mission for the *Palomino*. Instead, in May 2102, the *Phoenix A* will launch on its mission toward Barnard's star, and the *Phoenix B* will launch on its mission toward the Sirius star system.   Neither will rendezvous with the *Palomino* in May 2104.

So does the Phoenix *dock with the* Palomino *in 2104?*

From the *Phoenix A* crew's perspective – no. They were preparing for that mission, but it was scrubbed. When 2104 finally arrives, they are well on their way to Barnard's star, nowhere near Alpha Centauri or the *Palomino*. From the *Phoenix B* crew's perspective – yes. They remember that mission. They recall launching from Earth in June 2101, docking with the *Palomino* in May 2104, repairing her engines, and arriving back on Earth. But how could they have completed this mission since it was scrubbed in March 2100?

If the *Phoenix-Palomino* rescue mission does take place, then the success of the mission results in NASA's decision to scrub the mission, so that it does *not* take place. But if it does not take place, then the *Palomino* is still in danger, hence the mission *must* take place. This is a temporal paradox.

Recall that a paradox is something that seems contradictory. Some paradoxes are resolvable; there is no genuine contradiction. We have already seen that the twin paradox did not involve any genuine contradictions of objectively measurable quantities. It is my judgment that the temporal paradox above is a genuine contradiction. It is not something that is resolvable by appealing to a difference in perspective between reference frames. Either the *Palomino*'s engines get repaired in May 2104 or they don't. There is no third option here. And the crew of *Phoenix A* and *B* could theoretically go to Alpha Centauri at some later date to see if the *Palomino* is repaired or not. Either answer would contradict the experienced reality of one of the two *Phoenix* ships.

We know from the Lorentz transformation and the example above that faster-than-light speeds lead to the possibility of traveling backward in time. And we have seen that traveling backward in time can lead to a contradiction. But whatever leads to a contradiction is necessarily false. Therefore, the proposition that it is possible to travel faster than light, must be false. There are some caveats with this analysis, but the basic rule seems to be that any information-bearing system cannot travel backward in time, and therefore cannot travel faster than the speed of light. Let's analyze this proof using a common illustration.

The Grandfather Paradox

Suppose, for the sake of argument, that it was indeed possible to send a human being backward in time. That is, we set the past date we want to visit, step into the time machine, and are transported to the date we selected. In this scenario, it would be possible (however inadvisable) to meet your own paternal grandparents before they gave birth to your father. In principle, you could even prevent your grandparents from meeting each other in the first place. Now, given that you have prevented their meeting, it follows that they do not get married, and they do not have children. This means that your father was never born. But, if your father was never born, then neither were you. And if you were never born, how could you exist to then travel back in time and prevent your grandparents from having children?

If you prevent your grandparents from having children, then you cannot be born – in which case you cannot prevent your grandparents from having children since you do not exist. But if you do not exist, then you cannot prevent your grandparents from begetting your father who then begets you – in which case you do exist. So if you don't exist then you do. And if you do exist, then you don't. The possibility of traveling backward in time leads inexorably to the possibility of a genuine contradiction. Therefore, we must conclude that backward time travel for a person is not possible. And since faster-than-light travel leads to the possibility of backward time travel, it too must not be possible for a human being.

A similar argument can be made for any information-bearing system. A computer program could be set to activate a time machine at noon. The time machine then sends a message back in time to 11:55 to the computer, instructing the computer to abort and shut down before it can activate the time machine. If the message is sent, then the computer will shut down at 11:55 and will not activate the time machine at noon – in which case the message will not be sent. But if the message is not sent, then the computer will activate the time machine at noon, thereby sending the message back to 11:55 which shuts down the computer, so that the time machine never activates. Again, we have an impossible situation. If the message is sent, then the message cannot be sent. But if the message cannot be sent, then it must be sent – a contradiction. No human is present here, only autonomous, information-bearing machinery.

Sci-Fi Solutions to the Grandfather Paradox

I really enjoy science fiction movies and television shows – especially those that involve time travel into the past. When backward time travel is involved, the storyline must somehow deal with the possibility of a temporal paradox, such as those described above. Different authors have tried to resolve the temporal paradox in different ways. Some people have proposed that the real universe might employ one of these devices. Let's explore how each of these methods would attempt to resolve the grandfather paradox, and ask whether any of these could work for the real universe. (Please note that I am not necessarily recommending all the movies listed below. They are examples only, and some of them may contain objectionable material that is generally removed for network broadcasts.)

Perhaps the most common sci-fi solution to temporal paradox is the multiverse or "alternate timeline" method. Using this scheme, the time traveler who goes into the past and prevents his grandparents from meeting will experience no negative effects at all. He doesn't vanish, nor does the universe explode. Rather, he has effectively created a "new timeline" – one in which he will never be born. But since he was born in a different timeline from the one he now resides in, there is no contradiction. Since the moment he arrived in the past, he has effectively resided in a different universe from the one he was born into. Yes, he prevented his grandparents from meeting, thereby preventing his birth in universe B, but he is really from universe A where his grandparents did meet and he was born. If he were to time jump forward in time, it would be the future of universe B, in which he was never born.

The J.J. Abrams rebooted *Star Trek* movies (2009, 2012, 2016) make use of this device. With varying degrees of consistency, so do *Back to the Future Part II*, *Primer*, *Flight of the Navigator*, *Deja Vu* (2006), *Edge of Tomorrow, A Sound of Thunder, Galaxy Quest, Frequency, Star Trek: First Contact*, and *Superman* (1978). Several television series, including *Seven Days, Continuum*, and *Quantum Leap*, also allowed the past to be changed without any consequences for the time traveler.

The intellectual difficulty with the "alternative timeline" solution is that it requires an infinite number of universes, or at least one additional universe for every time travel event. And then we must ask which universe is the real one? Or at least, which is the universe in which I reside? This "solution" doesn't seem to provide any answer. Moreover, the act of traveling backward in time lands the traveler not in his own universe/timeline, but in a different one. The time traveler who goes back and prevents his grandparents from meeting did not really travel into his own past, but into an alternate universe where his parents are never born due to his interference. Hence, his journey was not truly time travel, but universe travel. This does alleviate the temporal paradox, but only at the expense of eliminating genuine time travel *into the real past of the same universe*.

Another proposed sci-fi solution is the "self-destruct" scenario. In this case, any time travel event that would lead to a genuine contradiction results in the destruction of the universe, or at least the destruction of the time traveler. This scenario was used (partially) in the movie *Back to the Future*, in which the time traveler, Marty McFly, accidentally prevents his parents from meeting. Marty is allowed a finite amount of time to remedy the situation before he starts to vanish.

This makes for a wonderfully entertaining movie. But it is not logically satisfying because it doesn't resolve the paradox. If Marty did prevent his parents from meeting, then he would never have been born and thus would not exist to travel back and prevent their meeting. Allowing him to simply vanish as a result of his mistake does not explain how he was able to make such a mistake if he never existed. Likewise, the destruction of the universe does not resolve the paradox. If the universe were destroyed, then the time traveler never existed to be able to cause a contradiction, resulting in the destruction. But if there is no contradiction, then what causes the destruction of the universe? So this solution doesn't really alleviate the paradox.

Another proposed solution is that backward time travel is only permitted in cases where the outcome is necessarily self-consistent. In this scenario, the time traveler would be unable to prevent his grandparents from meeting. In some versions of this scenario, the time traveler's actions that were intended to prevent his fate actually *cause* his fate, in an

Appointment in Samarra fashion. In other words, by attempting to prevent his grandparents from meeting, the time traveler inadvertently *causes* them to meet.

The Terminator movie partially makes use of this scenario. John Connor sends his best friend Kyle Reese back in time to protect John's mother Sarah. Kyle falls in love with Sarah, and becomes John's father. In this case, the time travel event is not only fully self-consistent, but is logically necessary. If John had not sent Kyle into the past, then John would never have been born, and would therefore not exist to send Kyle into the past.

Harry Potter and the Prisoner of Azkaban also makes use of this proposed solution. Harry observes someone perform a "Patronus Charm." He then travels backward in time and realizes that the person he observed performing the charm was actually himself. He is thus able to save himself by throwing a Patronus Charm, something he only knew he could do because he had previously seen it.

Escape from the Planet of the Apes and *Star Trek IV: The Voyage Home* also make use of this scenario. In the latter, the crew of the *Enterprise* gives the formula for transparent aluminum to the very person who "invented" it, which results in its discovery. In *Escape*, advanced apes from the future travel back to the present, where they produce an offspring. The baby grows, and his genetic material contributes to the creation of the advanced apes in the first place.

The advantage of this type of solution is that it is fully self-consistent. Only plotlines in which no paradox can occur are permitted. But, unfortunately, this answer would seem to disallow a true time machine in which the traveler had genuine freedom to tamper with the past. Are we really to believe that a time traveler would somehow not be able to prevent his grandparents from meeting? What would prevent him from doing so, assuming that his time machine is genuine? We must admit that it would be *possible*, however immoral or inadvisable, for the time traveler to simply shoot his grandparents before they gave birth to his parents. To suppose that they would magically become bullet-proof is far-fetched. So this seems an unlikely solution for the real universe.

The "self-correcting universe" is another option to remove paradoxes from sci-fi plots. In this scenario, if the time traveler prevents his grandparents from meeting when they originally did, the universe will adjust so that they meet at another time, this ensuring the time traveler's continued existence. Even if he succeeds in killing one of his ancestors, the universe will adjust so that some other person takes his place, and the right genes are selected so that the time traveler continues to live.

This option was used in the 2002 remake of the *The Time Machine*. Dr. Alexander Hartdegen is a scientist and inventor. When his fiancée is killed by a mugger, he devotes himself entirely to building a time machine, so that he might travel back and prevent the tragedy. He succeeds in saving her, only to have her killed by a horse and buggy. It seems that no matter what he does, the universe ensures that she will be killed in some fashion. Why can he not change the past? Later in the movie, he discovers the answer to the riddle. If his fiancée had lived, he would not have developed a time machine to save her; hence, the universe ensures that she cannot live so as to avoid a contradiction.

This attempted solution does have a certain elegance. But it doesn't eliminate the paradox because even a small change in the past is still a change. Though Dr. Hartdegen cannot save his fiancée from death, he does succeed in saving her from the mugger. Yet, he still remembers the incident. But if the mugging never happened, then how could he remember it? Moreover, there is no reasonable mechanism by which the universe could somehow "correct" intrusions in the timeline so that major historical events are never changed.

The conclusion of the whole matter would seem to be this: unrestrained time travel into the past is simply not possible because it inevitably leads to the possibility of contradiction. Even though not every time travel scenario would necessarily lead to a contradiction (as in the case of self-consistent outcomes), we must admit that it would at least be possible to cause a contradiction. And since that which leads to a contradiction must be false, we conclude that unrestrained time travel into the past is false. And since faster-than-light travel necessarily leads to unrestrained time travel into the past, we must conclude that faster-than-light travel is not possible for any information-bearing systems.

This proof of the impossibility of faster-than-light travel is based on the impossibility of changing past events, thereby causing a contradiction. To change a past event requires some type of information to be transmitted backward in time, so that actions can be taken to change the outcome from what it originally entailed. As such, the impossibility of faster-than-light (or backward in time) travel only applies to information-bearing systems, such as a person or a computer. A person who travels back in time would have knowledge of future events, and would therefore be able to alter events in a contradictory fashion. The same logic applies to a computer. It also applies to any sort of information-transmitting system, like a time telephone, in which the listener can hear instructions from the future.

But this proof does not apply to non-information-bearing systems. To my knowledge, there is nothing in relativistic physics that precludes the possibility of faster-than-light travel or backward-in-time travel for particles or systems that do not carry information. Such systems cannot be used to change the past or to cause any genuine contradictions, because they contain no information about the future. As a matter of fact, there are known phenomena that are faster than light, such as phase waves in a plasma. But these cannot be used to transmit information faster than light. So the possibility of faster-than-light travel for people or any information-carrying system seems to be eliminated on the basis that it leads to contradictions.

Chapter 11

The Relativistic Doppler Effect

When we derived the formula for time dilation, it was necessary to emphasize that moving clocks genuinely tick slower than stationary ones. They don't merely appear to be slower as viewed in a telescope, due to the progressively increasing distance that the light must traverse as the ship recedes. The latter is a mere optical effect, which also occurs. But it merely adds to the existing time dilation, so that what a person sees through a telescope is a combination of the effect of genuine time dilation plus the increasing time lag of light as the ship gains distance. But how do these effects combine? In this chapter we will explore what a person would actually see happening if he or she were to look through a powerful telescope into the window of a high-speed spaceship.

Let's first consider the classical limit, where time dilation is negligible. Let's suppose that Michael is in his train, moving directly away from Sarah at speed v, where v is much less than the speed of light. If Sarah were to watch Michael through a powerful telescope, his clocks would seem to tick ever-so-slightly slower than hers because the light from each tick of the clock takes a bit longer to reach Sarah than the previous tick did, due to the increasing distance. One consequence of this is the Doppler effect, which causes light to appear to have a different frequency or wavelength in a moving frame than it has in a stationary one.

Light is a traveling electromagnetic wave. A beam of light has peaks and troughs that travel through space. The frequency of light is the number of peaks that pass through a point in space in a given amount of time. So the frequency of light is much like the ticks of a clock. Since Sarah will see Michael's clock ticking slow, she will also see any light that Michael emits as having a reduced frequency from what Michael sees. This is a Doppler shift.

The Classical Doppler Effect

Suppose Michael shines a green laser pointer backward toward Sarah. A laser is simply light that has been created by a process that results in the light having a single frequency. This frequency is denoted by the lowercase Greek letter nu (ν). (Some textbooks use "f" for frequency since nu is more easily confused with "v." But nu is far more standard.) We can think of light as a traveling wave with peaks and troughs. These peaks are constantly passing by Sarah. The frequency is the number of peaks that pass by Sarah per second. Frequency is measured in hertz (Hz) which are the inverse of seconds (1 Hz = $1s^{-1}$). Visible light has a very high frequency, from 430 trillion Hz to 750 trillion Hz.

The frequency of light is related to its wavelength (the distance between two successive peaks) and its speed. Wavelength is denoted by the lowercase Greek letter lambda (λ), and of course the speed of light is c. From geometry, we can see that the relationship between these three quantities must be:

$$\lambda \nu = c \qquad (11.1)$$

So higher frequencies of light have smaller wavelengths, and vice versa. If Michael's green light has a wavelength of 532 nanometers (λ_s) (the most common value for a green laser pointer), then its frequency will be 563 trillion hertz. These numbers are the wavelength and frequency respectively in *his* reference frame. What will Sarah perceive to be the wavelength of this light (λ_0)?

The time between two successive peaks of the light is, by definition, simply the inverse of the frequency.

$$\Delta t = \frac{1}{\nu} = \frac{\lambda_S}{c}$$

But in this time, Michael has traveled a distance of $v\Delta t$. And the light has traveled a distance of $c\Delta t$ in the opposite direction. So Sarah will perceive the distance between two peaks as the sum of these two values:

$$\lambda_O = c\Delta t + v\Delta t = \lambda_S + \lambda_S \frac{v}{c}$$
$$\lambda_O = \lambda_S \left(1 + \frac{v}{c}\right) \tag{11.2}$$

This is the classical formula for the Doppler effect. Notice that it is sensitive to the sign of velocity. If Michael is moving away from Sarah, then v is positive and therefore Sarah will measure a larger wavelength (λ_O) for the light than Michael does (λ_S). The light will be shifted toward the longer, more "red" wavelengths; hence this is called "redshift." Conversely, if Michael is moving toward Sarah then v is negative, and Sarah will measure a shorter wavelength than Michael. The light will be shifted to a shorter wavelength, toward the blue end of the spectrum. Hence, the light will be "blueshifted."

We can see from the formula that the shift in wavelength between the source and the observer is quite small for ordinary velocities that are tiny compared to the speed of light. However, this formula also works for the shift in pitch associated with the velocity of sound, if we let c = the speed of sound. The speed of sound is only about 760 miles per hour at sea level. Achieving a significant fraction of that speed is quite attainable. A police car in pursuit may well travel faster than ten percent the speed of sound. And the human ear is quite sensitive to a change in sound frequencies. You have probably noticed the change in pitch as a fast car speeds by. So we have all experienced the Doppler effect with sound.

The classical formula for the Doppler effect (of light) ignores time dilation. The above proof assumes that Δt for Michael is the same as Δt for Sarah: an assumption that is increasingly inaccurate as velocity approaches the speed of light. The formula for the relativistic Doppler effect will be derived just as above, but will incorporate the fact that Michael's clock is time dilated relative to Sarah's clock.

The Relativistic Doppler Effect

As before, the time between two successive peaks from Michael's perspective is:

$$\Delta t' = \frac{1}{v} = \frac{\lambda_S}{c}$$

But Michael's clocks are running slow relative to Sarah's by the Lorentz factor. So Sarah will perceive the time between the emission of two peaks as:

$$\Delta t = \gamma \Delta t' = \frac{\gamma \lambda_S}{c}$$

In this time, the train has traveled a distance of $v\Delta t$. And the light has traveled a distance of $c\Delta t$. So the wavelength will be the sum of these two values:

$$\lambda_O = c\Delta t + v\Delta t = \gamma \lambda_S + \gamma \lambda_O \frac{v}{c}$$
$$\lambda_O = \lambda_S \gamma \left(1 + \frac{v}{c}\right)$$

We have our answer. And it is the same as the previous answer, except for the addition of the Lorentz factor. However, this expression can be further simplified by substituting the Lorentz factor and with some additional algebraic steps:

$$\lambda_O = \lambda_S \left(\frac{1 + \frac{v}{c}}{\sqrt{1 - \frac{v^2}{c^2}}}\right) = \lambda_S \left(\frac{\sqrt{\left(1 + \frac{v}{c}\right)\left(1 + \frac{v}{c}\right)}}{\sqrt{\left(1 + \frac{v}{c}\right)\left(1 - \frac{v}{c}\right)}}\right)$$

This simplifies to the usual way most textbooks express the formula:

$$\lambda_O = \lambda_S \sqrt{\frac{1 + \frac{v}{c}}{1 - \frac{v}{c}}} \tag{11.3}$$

This is the relativistic formula for the Doppler effect, when the light source is moving directly away from the observer at speed v, or directly toward the observer at speed $-v$. Since this formula includes the effect of time dilation, it works for all velocities up to the speed of light. Notice that when velocities are very small compared to light, this formula gives roughly the same answer as the classical Doppler effect formula. For example, if $v = 0.001c$, then both equations give a result of λ_O/λ_S of 0.001, which is the right answer.

Imagine a police officer pulling someone over for running a red light. The driver says, "But officer, at the speed I was traveling, the light was Doppler shifted so much that it looked green to me!" How fast would the car have to travel for red light to be Doppler shifted into green light? A typical wavelength for red light would be around 650 nanometers (or 6.5×10^{-7} meters) $= \lambda_S$. For green light, we will use 532 nanometers $= \lambda_O$ as before. Solving the Doppler effect formula for v, we find:

$$v = c \frac{\left(\dfrac{\lambda_O^2}{\lambda_S^2} - 1 \right)}{\left(\dfrac{\lambda_O^2}{\lambda_S^2} + 1 \right)}$$

This works out to 19.8% the speed of light, or about 133 million miles per hour! So it would probably be better to plead guilty to running the red light than to pay that speeding ticket. As with other relativistic phenomena, the Doppler effect only becomes noticeable to human vision at unattainably high speeds. However, with modern technology, it is possible to measure very small Doppler shifts in light.

Not So Easy to See

There is an additional reason why we do not readily see the Doppler effect in light. In the above thought experiment, we considered a light source that emits at only a single frequency/wavelength. But, aside from lasers, this is not common. Most objects that emit light produce a combination of wavelengths.

Our eye synthesizes the combination of wavelengths according to their relative strengths, and transmits the information to our brain which interprets the result as *color*. Color is a perception of the mind, not a property of photons. And so when we see "white light," this is not because any photon is "white." Rather, it means that we are seeing a combination of many wavelengths of light (short, medium, and long) in roughly equal proportions.

Furthermore, there are wavelengths of light that are too long (infrared) or too short (ultraviolet) for our eyes to detect. We simply cannot see these wavelengths, but they are there. So how does this pertain to the Doppler effect?

Imagine an object that emits an equal distribution of all wavelengths of light. It produces short wavelength light, long wavelength light, and so on in equal amounts – including wavelengths that are too short or too long for our eyes to see. What happens if this object moves away at some high fraction of the speed of light relative to us? Will it appear to turn red? No, its color will be unchanged. Here is why.

The object does experience a Doppler effect, so all its wavelengths will be longer than they were in the stationary frame. Short wavelengths that would have been interpreted as blue light have been shifted to long wavelengths associated with red light. So you might think it would appear red now. But in addition to this effect, wavelengths that were too short to see in the rest frame (ultraviolet) have been lengthened, so that they are now the wavelength of "blue" light. So the "blue" light that has been lost because it was shifted to red "has" been *replaced* by light that was originally ultraviolet, but is now shifted to blue. So the ratio of short wavelengths to long wavelengths has not changed. Therefore, the color remains the same.

In reality, most objects do emit a range of wavelengths of light, but they "peak" at a certain wavelength. So the observed color change due to the Doppler effect is intermediate between the laser and the uniform wavelength emitter scenarios above. At speeds that are a high fraction of the speed of light, the overall perceived color of most light emitters will indeed change. They will look redder for objects moving away at high speed, and bluer for objects that are approaching. But this is only noticeable at very high speeds, those that are a significant fraction of the speed of light.

Therefore, when you see stars of different colors in photographs, this is *not* due to the Doppler effect. Do not assume that the blue stars are approaching or that red stars are receding. Rather, stars come in different colors due to surface temperature. Blue stars are the hottest with surface temperatures in excess of 30,000 Kelvins. Red stars are the coolest with surface temperatures of around 4,000 Kelvins.

Astronomers can indeed detect the Doppler effect in stars, but not by eye. Stars contain "imperfections" in their frequency spectrum, where certain specific wavelengths of light are suppressed. Each of these wavelengths corresponds to a particular element at a particular temperature – a sort of atomic "fingerprint." And these wavelengths are known and always occur at exactly the same frequency in the rest frame for a given element. These subtle spectral lines can be detected by an instrument called a *spectroscope*. The spectroscope is able to measure the very slight wavelength shifts in these lines due to the Doppler effect from the star's radial motion. But the overall color of the star is not noticeably affected by velocity.

The Doppler Effect and Observed Time Dilation

Since the Doppler effect includes both (1) the time lag due to the increasing distance that light must travel between successive wavelength peaks, and (2) the actual time dilation of the moving clock relative to the stationary one, it represents the *apparent* slowing of time that a person would actually see. That is, if Sarah were to use a telescope to watch Michael's clocks tick, they would *appear* to her to be ticking slowly by the amount given in the Doppler effect formula, not the amount given by the time dilation formula. The former includes the optical effect of the increasing time lag of the light; the latter does not.

For example, suppose Michael's train is moving at 90% the speed of light. So $v/c = 0.9$. From the time dilation formula, his clocks will actually be ticking slower than Sarah's by a factor of 2.29. But if Sarah were to watch Michael's clocks through a telescope, they would *appear* to be ticking 4.36 times slower than her clocks (from the Doppler effect formula). The factor of 2.29 comes from genuine time dilation; the additional 2.07 comes from the fact that Michael is at a progressively greater distance from Sarah during each tick of the clock, and so the time it takes for light to cover the distance increases by 2.07 seconds for every tick. (At near light speed, roughly half of the observed slowing of clocks is due to real time dilation, and half is the optical effect of the increasing time lag.)

What if Michael *approaches* Sarah at $0.9c$? In this case, his clocks will still be ticking slower than hers by a factor of 2.29, but what will she see through her telescope? Using the relativistic Doppler effect formula, she will see Michael's clocks appear to be ticking 4.36 times *faster* than her own. For each second that elapses on Sarah's clock, Michael has traveled closer to Sarah by $0.9c$; so the light from his clock takes 0.9 less seconds to reach her, as if each second takes only 0.1 seconds on Sarah's clock. So if Michael were to experience no time dilation, Sarah would see his clock ticking *ten* times faster than hers. But since his clocks actually tick slower than hers by 2.29, she sees the apparent ticking of his clocks to be 10/2.29 = 4.36 times faster than hers.

Transverse Doppler Effect

Another weird phenomenon in relativity that does not occur in classical physics is the transverse Doppler effect. In classical physics, the Doppler effect is the stretching or compression of wavelength as the distance between emitter and receiver changes. In relativistic physics, the Doppler effect is due partly to this changing distance, but also includes the effect of time dilation. What if we consider the case where something is in motion, but the distance does not change? In this case, any Doppler effect will be due entirely to time dilation.

Consider a spaceship moving in a circle around the Earth at some high speed v. The distance between the ship and the Earth is constant. Therefore, the time it takes for light to travel from the ship to Earth is also constant. Yet, there will still be a Doppler shift due to time dilation.

Let λ_S represent the wavelength of the light as observed by the person on the ship. The light is aimed so that it travels from the ship to the Earth. What wavelength (λ_O) will the observer on Earth see? Since the distance remains constant, the only relevant effect is time dilation – the formula we derived earlier. So the frequency that the passenger on the ship observes will be higher than the frequency the observer on Earth sees by the Lorentz factor:

$$\nu_O = \frac{\nu_S}{\gamma}$$

Since the frequency is related to wavelength by $\lambda\nu = c$, we have:

$$\frac{c}{\lambda_O} = \frac{c}{\gamma\lambda_S}$$

$$\lambda_O = \gamma\lambda_S \qquad (11.4)$$

We see that there is a change in the measured wavelength, even though the distance between source and observer is constant. There, the transverse Doppler effect can only be explained in the context of relativity, due to the effect of genuine time dilation.

Note that this scenario is not symmetric in this instance. That is, if an observer on Earth sent a light beam to the ship that circles the Earth at high speed, the observer on the ship would see the light blueshifted – not redshifted. The reason for the lack of symmetry is that this observer is not in an inertial reference frame. His velocity is constantly changing direction, even though the magnitude is constant.

However, if we were to redo this experiment so that both observers were inertial, traveling at constant velocity in opposite directions, then the situation would be symmetric. At the moment of closest approach, when each observer sees the other at a right angle relative to the direction of motion, each observer will see the other's light redshifted.

It may seem paradoxical at first that each observer will see the other observer redshifted. But there are no genuine contradictions, because each observer uses his own coordinate system to measure lengths and times. And the two coordinate systems will differ, thanks to the Lorentz transformation. This is analogous to the solution to the "trapping the train" thought experiment.

Relativistic Acceleration

The Doppler shift formula allows us to solve some relativistic problems that would be difficult to solve by other methods. One of these concerns the time dilation experienced in an accelerating frame of reference. Accelerating frames are non-inertial, but relativity still allows us to examine these frames from the perspective of an inertial frame. One interesting phenomenon that occurs in an accelerated frame is a type of time dilation that we have not yet discussed – one that is due to the relativity of simultaneity.

Recall from the "trapping the train" experiment that when one observer computes that two events are simultaneous, another observer with a different velocity will not, in general, compute that the same two events are simultaneous. If we take the moving observer's motion to be in the positive x direction, then he will see the event with a greater value for the x-coordinate occur before the other event.

So what happens to an observer that is accelerating – increasing his velocity in the positive x direction? Events with a very positive x coordinate will seem to occur progressively earlier, whereas events with a negative x coordinate will occur progressively later.   Therefore, the universe in front of the accelerating observer will seem to be sped-up due to this effect, whereas the universe behind the observer will appear slowed-down. What is even more astonishing is that this effect will occur even within his own ship. We can use the Doppler shift formula to compute the effect of time dilation due to the acceleration of the ship.

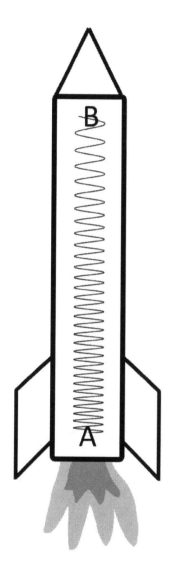

Imagine a very tall rocket ship with passengers near the bottom at A and passengers near the top at B. The rocket is deep in space, far away from any noticeable effects of gravity, and is stationary (for the moment) relative to some inertial onlooker. The rocket ignites its engines and begins accelerating upward in the positive y direction. Since the velocity of the ship is constantly changing, the passengers at the bottom of the ship will see events above them (including the passengers at B) occur progressively earlier, as if slightly   sped-up. Conversely, the passengers near the top at B will observe the passengers at A moving slowly, since events in the negative y direction occur progressively later.

So the passengers near the bottom of an accelerating rocket will experience time dilation relative to the passengers at the top. But how much time dilation occurs? The easiest way to answer this question is to use the Doppler shift formula. At the moment the rocket starts its engines, imagine that a passenger at A turns on his flashlight and aims it upward to a passenger at B. The light leaves the flashlight with some wavelength λ_s. It travels a distance of h, the distance between A and B.

When the light arrives at B, what will the passengers see as its wavelength? It will no longer be λ_s because in the time it took the light to traverse the distance h, the rocket has changed its velocity due to acceleration. Passengers at B now have some upward velocity that is greater than the velocity of the flashlight when the light was emitted. So the light will be slightly redshifted.

To make this problem more tractable, we will use some approximations. First, let's assume that the acceleration and the height of the rocket are both relativity small, so that the change in velocity of the rocket between when the light was emitted at A and received at B is much smaller than the speed of light. This means that we can neglect the small change in the position of B as the light reaches it. So as an approximation, the time it takes the light to traverse the distance h is:

$$\Delta t = \frac{h}{c}$$

Since acceleration is the change of velocity divided by the change in time, we have:

$$a = \frac{v}{\Delta t}$$

And so, the velocity of the passengers at B *when the light is received* relative to the flashlight *at the time the light was emitted* will be:

$$v = a\Delta t = \frac{ah}{c}$$

Since we have taken acceleration and height to be small, it follows that the velocity v will be small. Therefore, we can use the classical formula for the Doppler shift, which make the algebra significantly easier. So the wavelength that the observers at B will see λ_0 will be:

$$\lambda_O = \lambda_S \left(1 + \frac{v}{c}\right)$$

Substituting our formula for velocity, we end up with the answer:

$$\lambda_O = \lambda_S \left(1 + \frac{ah}{c^2}\right) \tag{11.5}$$

Thus, passengers at the top of the rocket will see light from the bottom of the rocket as slightly redshifted. But notice that the distance between the bottom and top of the rocket does not change! Passengers at A are neither approaching nor receding from passengers at B; nonetheless there is a Doppler shift.

Recall that the (relativistic) Doppler effect includes both apparent time dilation due to the changing distance between the light source and the observer, and also the effect of genuine time dilation. But in the above scenario, the distance between passengers at A and those at B does not change. Therefore, from the passengers' point of view, any Doppler shift must be due entirely to time dilation. After all, the bottom of the ship is neither approaching nor receding from the top, as far as the passengers are concerned.

In an accelerating frame, clocks at lower positions in the frame tick slower than clocks at higher positions in the frame. The difference in rate is given by the Doppler shift formula above. So if two clocks are in the same reference frame (accelerating at *a*), and clock B is higher (in the direction of acceleration) than clock A by *h*, then the time dilation between these two clocks as measured by someone in that frame will be:

$$\frac{\Delta T_B}{\Delta T_A} = 1 + \frac{ah}{c^2} \tag{11.6}$$

Keep in mind that this is an approximation for low values of both *a* and *h*. And since both *a* and *h* are in the numerator of the right-hand term, and since c^2 (which is quite large) is in the denominator, it follows that the time dilation will be quite small under the conditions we selected.

Time dilation due to acceleration will be small for ordinary velocities. Yet, this is a very significant result, because it will lead to another type of time dilation once we include the effects of gravity. We will revisit this topic in a later chapter. In the meantime, there are two additional strange effects that become apparent as we approach the speed of light: Lorentz aberration and Lorentz beaming. These will have a distorting effect on the appearance of the universe when an observer approaches the speed of light and will be the subject of the next chapter.

Chapter 12

Lorentz Aberration and Lorentz Beaming

The constancy of the speed of light in all inertial reference frames implies time dilation, length contraction, and the relativity of simultaneity. These effects are expressed in the Lorentz transformation, which allows us to convert the coordinates of any event as measured by one observer into the coordinate system of another observer moving at a different velocity. This transformation implies that the way the universe looks when traveling at a very high speed is quite strange and different from our expectations. As a spaceship accelerates, the universe appears to contract in the direction of motion. Clocks in the universe seem to run progressively slower, yet paradoxically, observers in the universe see clocks on the ship running slower. But there are two additional effects that we have not yet discovered.

The first of these effects concerns the position of objects in the universe. So far, we have primarily dealt with only the x and time dimensions. And although the y and z dimensions are not affected by motion along the x-axis (recall $y = y'$, and $z = z'$), nonetheless, when we begin analyzing angles as measured by observers in different reference frames, some startling effects become apparent. Perhaps most significant among these is Lorentz aberration.

Lorentz Aberration

On a clear, summer night, Sarah is outside looking through a telescope at a star that happens to be directly overhead. Let's imagine that the star is

100 light-years away, though the distance isn't really critical to our discussion. If we take "up" to be the positive y-direction, then the star has spatial coordinates $x = 0$, $y = 100$, $z = 0$ light-years. Light from the star travels through space, directly down the barrel of the telescope, and into Sarah's right eye.

Michael also enjoys astronomy, but he prefers to set up his telescope on his high-speed train. Let us suppose that he is traveling in the positive x direction ("to the right") and passes by Sarah at the very moment she is looking through her telescope. To be precise, let's suppose that Michael passes by Sarah at the exact moment that a particle of light from the star enters the top of her telescope. The question is this: Will Michael also observe the same star to be directly overhead?

First let's analyze the situation geometrically. Supposing that Michael's telescope is also pointed directly overhead, will he see the same star that Sarah sees? If a photon from that star enters his telescope at the exact moment he passes by Sarah, will it reach his eye? No. Although the photon is traveling directly downward in the negative y direction, Michael is moving at high speed in the positive x direction. From geometry, we can see that the photon will collide with the left side of the telescope before ever reaching the base.

In order to see the same star that Sarah sees, Michael will have to tilt his telescope forward a bit, so that it is pointed in the positive x and positive y direction. With the telescope tilted, the light beam can enter the top and travel directly downward while the telescope moves forward. Only then will the light reach the base of the telescope and enter Michael's eye. Clearly, the faster Michael travels, the more he will have to tilt the telescope forward in order to see the star. So whereas the star is directly overhead for Sarah, the star is above *and in front of* Michael. We can also see that this effect must happen by consulting the Lorentz transformation. This will tell us how much Michael's telescope must be tilted in order to see the star.

Derivation of Lorentz Aberration for a Special Case Where $\theta = 90°$

Sarah looks at a star at distance r (100 light-years). If Michael travels at speed v relative to Sarah, and Sarah sees the star directly overhead ($\theta = 90°$) with coordinates $x = 0$, $y = r$, $z = 0$, what will the coordinates be for Michael? And at what angle will he have to tilt his telescope to see the star? Since the speed of light is one light-year per year, the time at which the light is emitted by the star from Sarah's perspective is $t = -r/c$ (which is -100 years). This must be the case since she sees the light at $t = 0$. So we have $x = 0$, $y = r$, $z = 0$, $t = -r/c$. Using the Lorentz transformation (8.2), what are x', y', z', and t'?

$$x' = \gamma(x - vt) = \gamma vr/c$$
$$y' = y = r$$
$$z' = z = 0$$
$$t' = \gamma\left(t - \frac{vx}{c^2}\right) = -\gamma r/c$$

Supposing that Michael travels at 0.9c, then this works out to: $x' = 206$, $y' = 100$, $z' = 0$, $t' = -229$.

From geometry, we can see that the angle θ' that Michael tilts his telescope away from vertical will be given by:

$$\tan \theta' = y'/x'$$

At 0.9c this works out to an angle of 26 degrees above horizontal. But in general, the angle is given by substituting our expressions for x' and y' into the above equation:

$$\tan \theta' = \frac{c}{\gamma v} \tag{12.1}$$

So for a given velocity, we can compute how much Michael will need to tilt his telescope in order to see a star that was directly overhead from Sarah's point of view. From her perspective, Michael is in motion and therefore has moved in the time between when the light enters his telescope and when it hits his eye. So naturally he must tip the telescope forward a bit. But how are we to explain the situation from Michael's reference frame?

From his frame, Michael is stationary, and Sarah and the rest of the Earth are moving in the negative x direction at some high-speed v. He perceives the speed of light to be constant relative to himself. So how does he explain the tilt of his telescope? Since he is not moving from his point of view, he must point the telescope at angle θ because that is actually where the star is! As far as Michael is concerned, the star isn't directly overhead at all, but has a large positive x-coordinate. But then why does Sarah think the star is directly overhead?

From Michael's perspective, Sarah is moving in the negative x direction at high speed. If her telescope were pointed directly at the star, at angle θ, the light would not reach her eye because the telescope is in motion. The light beam would strike the right side of the telescope. Therefore, Sarah must tilt her telescope counterclockwise in order for the light to avoid collision with the telescope sides, and ultimately reach the eyepiece. Due to her motion, Sarah must tilt her telescope so that it is vertical, even though the star is actually at angle θ.

So the reasoning is exactly the same as before. Each observer imagines that he or she has the telescope pointed in the *actual* direction of the star. Each sees the other person as moving, and therefore the other observer must tilt his or her telescope at an angle away from the star in order for the light beam to reach the eyepiece.

The Lorentz Aberration Angle

Above, we considered only the special case where Sarah sees the star directly overhead. What about stars at other angles? In general, Michael will not see them at the same angle as Sarah. Rather, he will have to tilt his telescope clockwise relative to Sarah in order to see the star in his eyepiece. To compute how much, we consult the Lorentz transformation (8.2). We will limit our consideration to stars in the x-y plane (the aberration angle works exactly the same in the x-z plane as in the x-y plane). Consider a star at distance r away from Earth relative to Sarah. Its angle relative to Sarah is θ. We will use the standard practice of taking the angle relative to horizontal. (So a star directly overhead would have $\theta = 90°$.) Then the coordinates of the star will be:

$$x = r \cos \theta$$
$$y = r \sin \theta$$
$$z = 0$$
$$t = -r/c$$

Substituting these values into the Lorentz transformation (8.2), we get the primed coordinates:

$$x' = \gamma(x - vt) = \gamma\left(r\cos\theta + \frac{vr}{c}\right)$$

$$y' = y = r\sin\theta$$

$$z' = z = 0$$

$$t' = \gamma\left(t - \frac{vx}{c^2}\right) = \gamma\left(-\frac{r}{c} - \frac{vr\cos\theta}{c^2}\right)$$

From trigonometry, the tangent of θ' must equal y'/x'. So we have:

$$\tan\theta' = \frac{y'}{x'} = \frac{r\sin\theta}{\gamma\left(r\cos\theta + \frac{vr}{c}\right)}$$

$$\tan\theta' = \frac{\sin\theta}{\gamma\left(\cos\theta + \frac{v}{c}\right)} \tag{12.2}$$

The above formula will allow us to convert the angle of any event from the coordinates of one observer to any other observer. The general trend is illustrated in the accompanying figure. Those objects that appear directly above (or directly below) the stationary observer will appear shifted forward (in the direction of motion) to the moving observer. This is the case with stars 3 and 7 in the figure.

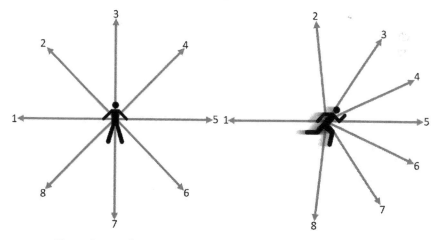

Therefore, if you were in a spacecraft and accelerating to progressively greater speeds, the universe would not only contract in the direction of motion, but the stars would begin to shift forward, as if gathering toward the front of your ship. Essentially, angles in the forward direction of motion contract (such as the angle between stars 4 and 5), while angles in the backward direction expand (such as the angle between stars 1 and 2). This causes objects in the forward direction to appear smaller and objects in the backward direction to appear larger than they would to the stationary observer.

Lorentz Beaming

The effect of Lorentz aberration naturally leads to the related phenomenon of Lorentz beaming. The above derivation of Lorentz aberration is symmetric: meaning it doesn't matter whether the light is incoming from distant stars to the observer, or outgoing from the observer to distant stars. The aberration angle is same.

So imagine the illustration above, but replace the observer with a light bulb. The arrows now represent outgoing light rather than incoming starlight. A stationary light bulb emits light equally in all directions (neglecting the base), and this is represented by the arrows on the left panel of the figure. They are equally spaced in direction, indicating that as much light energy is emitted toward the left as toward the right.

In the right panel, the light bulb is now moving at high velocity to the right relative to some onlooker. The emission of light is no longer evenly distributed in all directions. Rather, the light bulb now emits more energy to the right – in the direction of motion. The faster the bulb moves, the more its light becomes concentrated in the direction of motion due to Lorentz aberration. Therefore, a person standing in front (to the right) of the bulb would see more light than a person standing to the left at equal distance. The bulb begins to act more like a flashlight beam, where the energy is directed forward rather than evenly spaced. This effect is called Lorentz beaming.

Suppose we had two light sources of equal brightness (when stationary) at an equal distance from an observer. But one light source is moving toward the observer, while the other is moving away. As a result of Lorentz beaming, the light source moving toward the observer will appear somewhat brighter than the one moving away. And from the Doppler shift formula we derived in the previous chapter, the oncoming light source will be blueshifted, and the receding object will be redshifted.

As with other relativistic effects, Lorentz beaming only becomes apparent when velocities are a significant fraction of the speed of light. It is negligible for stars in our night sky, which move very slowly compared to the speed of light. But there are objects in the universe that exhibit very strong Lorentz beaming. Astrophysical jets are collimated streams of matter that are ejected from a compact object. The details of how such jets are generated is the subject of some debate. But astronomers have discovered many such jets, often at the core of a galaxy such as a quasar. In many cases the material in astrophysical jets moves at a substantial fraction of the speed of light. Therefore, relativistic effects become quite noticeable.

Interestingly, most relativistic jets that have been so far discovered are aimed in our general direction. That is, they are directed roughly toward our solar system. That might seem strange until we consider Lorentz beaming. Those jets directed toward us will appear much brighter than those directed away. Therefore, the Earth-directed jets are much easier to discover.

The Appearance of the Universe at Near-Light Speed

Combining all the relativistic effects we have so far discovered, what would the universe look like if we were to travel through it at nearly the speed of light? As our spacecraft gains speed relative to the nearby stars, they will begin to contract in the direction of motion. So a star directly in front or behind our ship will be much closer than it was when we were stationary. Indeed, the entire universe will seem to compress in the direction in which we are accelerating. Stars will no longer appear exactly spherical. They will be squashed in the direction we are accelerating.

Furthermore, when we watch events in front of us through a telescope, they will seem to be "sped up," as if fast-forwarding through a movie. Events behind us will seem slowed down, as if watching a movie in slow-motion. Events directly beside us will also appear slowed down. But of course, we know that the apparent speeding up of events in front of us is because they are moving toward us, and so the light takes less time to reach us with each passing second. When we subtract this optical effect, we find that all clocks in the local universe are actually running slow compared to our clock on the ship.

As we gain speed, the stars will seem to shift forward. The star that was directly overhead when we were stationary is now above *and in front of* us. It looks as if the universe in our forward direction is shrinking, as if we were zooming out through a camera. Conversely, the universe behind us seems to be expanding, as if we were zooming in. Yet, it becomes increasingly difficult to see the stars behind us; although they are magnified, they are becoming quite dim, and now appear redder than they did when we were stationary. Conversely, the stars in front of our ship appear much brighter and bluer than they did before.

If we were to continue to accelerate, these effects would become even more pronounced. It would seem like the entire universe were becoming concentrated into a small, very bright blue beam directly in front of us. It would be as if we had tunnel vision. Objects to our side or behind us would be very dark, and nearly impossible to see. The universe would become increasingly compressed in our direction of motion. And so we could travel past stars in a relatively short period of time, because the distance between stars (in the direction of motion) has become very short.

It may seem strange. But we have mathematically proved that these effects must occur if indeed the two postulates of relativity are correct. So far, our analysis has focused on the effects of space and time, and therefore the instruments used to measure space and time: tape measures and clocks. But does traveling at nearly the speed of light have any strange effects on other properties, such as mass, or energy?

Chapter 13

Energy, Mass, and Momentum

We have seen that some seemingly strange effects occur with length and time as an object approaches the speed of light. In this chapter we will find that mass and momentum also behave in some unexpected ways. The examination of the relationship between mass and energy in light of the physics of Einstein led to one of the most remarkable discoveries of the twentieth century: something we will find in the next chapter. But we must first discover how mass behaves as speed increases.

Mass is a difficult quantity to define in a non-circular way. Basically, mass is a resistance to a change in motion. The more mass an object has, the harder it is to push on it and make it move. Under Earth's gravity, mass produces weight. A more massive object is heavier than a less massive object. But this trend does not work in the depths of space, far from any source of gravity. In deep space, a massive object has no weight at all. But it still has the same mass; therefore, it would still be hard to make it move very much by pushing on it.

Energy is the ability of a system to perform work; and work is the ability to change an object's state of motion. Basically, a thing possesses energy when it is moving or when it can make something else move. There are two broad categories of energy: kinetic energy and potential energy. Kinetic energy is the energy possessed by a system due to its motion. A baseball bat has kinetic energy when it is moving. And it has the capacity to change the baseball's state of motion upon striking it. Potential energy is any form of energy that is *not* due to motion. A battery has potential energy, because when connected to a motor it can make the motor move, thereby changing its state of motion. Kinetic energy is energy of motion; whereas potential energy is "stored."

In classical physics, the formula for the kinetic energy of a point mass is given as:

$$E = \frac{1}{2}mv^2$$

Here, m is the mass of the moving object, and v is its velocity. This formula works very well as an approximation for objects whose velocity is much less than the speed of light. But we will find that it is not correct for high velocities. In the next chapter we will seek to discover the relativistic formula for energy that works at all velocities. For now, we will use the above equation as a low-velocity approximation.

Energy is a *conserved quantity*. This means that in an isolated system (one in which no energy or mass is allowed to escape or enter from the outside), the total energy remains the same at all times. Energy cannot be created or destroyed. This is called the law of conservation of energy, and is also the first law of thermodynamics. It is a very important principle in science, and we will use this principle in some of our derivations below. Energy can be transformed from one type to another, and it can be moved from one location to another. But the energy of the universe cannot be increased or decreased.

A distinct but related concept is momentum. Momentum is a combination of velocity and mass; the more mass and velocity an object possesses, the more momentum it has. A semi-truck moving at 50 miles per hour has more momentum than a fly at the same speed, because it has more mass. Momentum is denoted by a lowercase p. In classical mechanics, the momentum of an object is simply the product of its velocity and mass.

$$\boldsymbol{p} = m\boldsymbol{v} \tag{13.1}$$

Whether or not this formula remains true at high velocities remains to be seen. But this is at least a good approximation at low velocities.

Vectors

In a 3-dimensional universe, both velocity and momentum are *vectors*. A vector is any quantity that has both a magnitude and a direction. A car might have a velocity of 30 miles per hour northward. The 30 miles per hour is the magnitude of velocity, and the "northward" is the direction of the velocity. The magnitude of velocity is defined as "speed."

We may express the value of a vector either by giving its total magnitude and then its direction, or by giving its x, y, z components. For example, suppose a car were traveling at 50 miles per hour in a northeast direction. Defining north to be $\theta = 0°$, and east to be $\theta = 90°$, we could express this as: velocity v is 50 miles per hour, $\theta = 45°$. Alternatively, we could express each component of velocity: the velocity v is [35.36, 35.36, 0] miles per hour, or $v_x = 35.36$ mi/hr, $v_y = 35.36$ mi/hr, $v_z = 0$ mi/hr. The subscripts x, y, z are used to indicate which component of velocity is addressed. Any of these notations is legitimate, and we can use trigonometry to convert from one to the other, as shown for two dimensions in the figure. The details are not necessary for our purposes.

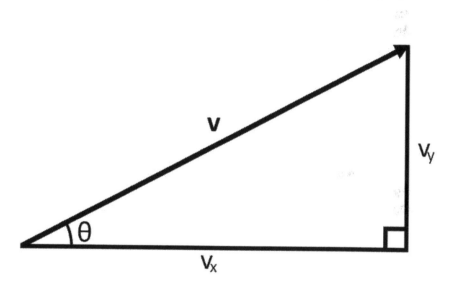

When in a printed book, vectors are usually indicated in **bold** letters, but when handwritten, they have a small arrow over the letter as in: $\vec{v}$. Quantities that do not have a direction, but merely a magnitude are called *scalars*. Mass and energy are scalar quantities. When we multiply a vector times a scalar, only the magnitude of the vector changes, not its direction. Each component of the vector is multiplied by the scalar. Thus, from the classical equation for momentum, the direction of the momentum is always the same as the direction of the velocity of the object.

Momentum is defined in terms of force. Specifically, the change in momentum of an object during time interval Δt is defined to be the net force $\boldsymbol{F}$ on that object multiplied by the time interval:

$$\Delta \boldsymbol{p} = \boldsymbol{F}\Delta t \qquad (13.2)$$

Note that force is a vector quantity because you can push an object north, south, up, down, east, west, or any combination thereof. But the time interval is a scalar. So the *change* in momentum will always have the same direction as the force being applied on the object. Since the above equation is the definition of momentum, it is necessarily true at all velocities.

Likewise, energy is defined in terms of force. Specifically, the change in energy is equal to the force applied on an object multiplied by the distance over which the force is applied. In one-dimension, this is expressed as follows:

$$\Delta E = F\Delta x \qquad (13.3)$$

Here F is the net force applied in the x-direction (since it is only in one dimension, it is scalar here), and Δx is the distance over which the force is applied.

Mass

When it comes to the physics of Einstein, there is some disagreement among physicists as to how exactly we should define mass, and how it relates to momentum. One camp suggests that mass should be defined as an object's resistance to motion in the limit as velocity goes to zero. In other words, mass should always be evaluated in the reference frame in which the object is stationary. Using this system, all observers, regardless of their velocity will compute the same mass for a given object, because they will all evaluate the mass in its reference frame rather than their own. By this definition, the mass of an object does not depend on its velocity. However, using this system, the classical formula $\boldsymbol{p} = m\boldsymbol{v}$ will *not* be true at high speeds, and instead we will need to derive the relativistic formula for momentum. This is the more common position.

Alternatively, we can *define* mass on the basis of the classical momentum formula (13.2), such that $m = \boldsymbol{p}/\boldsymbol{v}$. In this scenario, the classical momentum formula is necessarily correct at all speeds – by definition. And since it relates directly to momentum under this definition, mass is a conserved quantity. That is, the mass of an isolated system cannot be increased or decreased. However, we may *not* assume that mass is unaffected by motion. That is, we may not arbitrarily assume that what one observer measures as the mass of an object is what other observers (moving at a different velocity) measure as the mass of that object.

Note that either option has its weirdness. It is merely a question of *where* we want the weirdness to be. If we take mass to be that which is measured in its stationary frame, then mass is unchanged at all velocities, but momentum behaves strangely and no longer obeys the rule $p = mv$. On the other hand, if we take $p = mv$ to define the relationship between mass and momentum, then we find that momentum is well-behaved at all speeds, but mass isn't. Although this second position is less common, I believe it has tremendous advantages, particularly when it comes to understanding the underlying physics. And so this is the position I will take in this chapter. We will see at the end that it is easy to convert from one system to the other.

Defining $p = mv$, the mass of a moving object is not necessarily the mass it has *when it is at rest*. The mass that an object has at rest we will simply call "rest mass" and designate this by the letter m with a subscript of zero: m_0. Just as the length of a moving train is not the same as its length when stationary, we are going to find that the mass of a moving object is not exactly the same as its rest mass. In general, the mass of a moving object will be higher than its rest mass, for all positive velocities: $m > m_0$. Amazingly, we are going to find that objects get "heavier" as they approach the speed of light. But what is the formula that relates the actual mass of a moving object to its rest mass?

Derivation of Relativistic Mass

Imagine two particles, each with a rest mass of m_0. Suppose that these two particles are moving at identical speed (w) but in opposite directions, relative to observer O (upper left panel of the figure). From this observer's perspective, the mass of each particle will be m_1, which is not exactly the same as m_0. Suppose the two particles are "sticky" so that when they collide, they combine to form a single object of mass M_0 (upper right panel). From symmetry, it is obvious that particle M_0 will be stationary, since the two particles had equal mass and traveled at equal speeds in opposite directions.

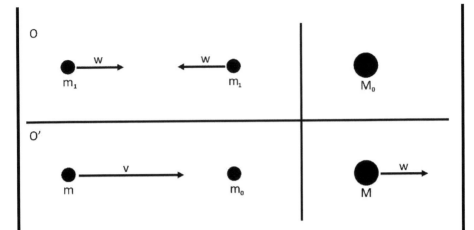

Now imagine the same scenario as viewed by observer O'. This observer moves at the same velocity as the particle on the right side, so he sees it as stationary and with mass m_0, which is the rest mass of the particle (lower left panel). The left-hand particle will be moving to the right at speed v, with mass m. We might at first be tempted to think that $v=2w$, but this falsely assumes the Galilean velocity addition theorem. The left-hand particle moves at speed w relative to O, who moves at speed w relative to O'. So we use the relativistic velocity addition formula (8.4) we derived in chapter 8 to find the speed of the left-hand particle (v) relative to observer O'.

$$v = \frac{2w}{1 + \dfrac{w^2}{c^2}}$$

This formula gives v in terms of w, but it will be more useful for us to have w in terms of v. So we solve this equation for w, first by rearranging the terms into a quadratic equation:

$$\frac{v}{c^2}w^2 - 2w + v = 0$$

We then use the quadratic formula to solve this for w, and we select the negative root since we can see from geometry that $w<v$. This yields:

$$w = \frac{c^2}{v}\left(1 - \sqrt{1 - \frac{v^2}{c^2}}\right)$$

Since observer O' moves at speed w relative to O, and since the combined mass (M_0) after the collision is stationary relative to O, it follows that the combined mass will move at speed w relative to observer O (lower right panel) and its mass will be M. Since relativistic mass is conserved, it follows that the total mass before the collision must be equal to the total mass after the collision (in either reference frame). In terms of the O' frame this means:

$$M = m + m_0$$

And since momentum is conserved, the total momentum before the collision must be equal to the total momentum after the collision. From the O' perspective, this means:

$$mv = Mw$$

Now we combine these last three equations, and eliminate M and w in order to get m in terms of m_0 and v. We start by substituting our formulas for M and w into the conservation of momentum formula. This yields:

$$mv = (m + m_0)\frac{c^2}{v}\left(1 - \sqrt{1 - \frac{v^2}{c^2}}\right)$$

Solving this for m and simplifying gives:

$$m = \frac{\sqrt{1 - \frac{v^2}{c^2}} - 1}{1 - \frac{v^2}{c^2} - \sqrt{1 - \frac{v^2}{c^2}}}m_0$$

The denominator of the right-hand side can be further simplified giving:

$$m = \frac{\sqrt{1 - \frac{v^2}{c^2}} - 1}{\left(\sqrt{1 - \frac{v^2}{c^2}} - 1\right)\sqrt{1 - \frac{v^2}{c^2}}} m_0$$

The numerator cancels with the first term in the denominator, giving our solution:

$$m = \frac{m_0}{\sqrt{1 - \frac{v^2}{c^2}}}$$

Recall from 5.2 that the that Lorentz factor is $\gamma = 1/\sqrt{1 - v^2/c^2}$. So substituting this into the formula, we find:

$$\boxed{m = \gamma m_0} \tag{13.4}$$

Remarkably, mass increases with velocity in exactly the same way that time dilates! The faster something moves, the "heavier" it becomes relative to a stationary observer.

As with length contraction and time dilation, the person moving at high speed does not feel the effect of his own increase in mass. From his perspective, he is stationary, and his mass remains the rest mass. But an outside observer would find that the high-speed traveler has increased in mass by the Lorentz factor.

As we have come to expect, this increase in relativistic mass with velocity is not at all noticeable in our day to day experiences because the effect is so small at "ordinary" speeds. An object would have to be accelerated to 14% the speed of light in order to gain a mere 1% increase in mass relative to its rest mass. But the effect becomes very significant at high speed. At 90% the speed of light, the mass of an object has more than doubled from its stationary value. At 99% the speed of light, the object is 7 times as "heavy" as its stationary mass!

Implications of Relativistic Mass

So a moving train will be somewhat "heavier" (more massive) than a stationary train of identical construction (identical rest mass). The faster an object moves, the more massive it becomes. This does not mean that the object becomes larger in volume. On the contrary, the *volume* decreases. Recall that the length of a moving object is contracted in the direction of motion from its stationary value, while height and width are unaffected. So an outside observer would find the moving train to have a higher density than a passenger on the train would measure.

Suppose we take a mass m_0 and apply a force to it so that it now moves at velocity v. The mass will now be m, something a bit larger than m_0 by the Lorentz factor. If we now apply the same force again to the moving mass, will its velocity go to $2v$? No, because the object is now a bit heavier than it was before. So applying the same force as before will not result in the same change in velocity as before, but something less. The new velocity will be less than $2v$.

Of course, at "ordinary" speeds much less than c, the change in mass is negligible, and so the new velocity will be essentially $2v$. But when the velocities are high, the change in mass becomes substantial, and the new velocity will be noticeably less than $2v$. As we continue to apply the same force to the object, its speed continues to go up, but its mass continues to increase.   With each increase in mass, it becomes progressively harder to accelerate the object further. At 99.99% the speed of light, the object is 70 times more massive than it was when stationary. And applying a force just doesn't produce very much acceleration.[13]

If we consider the case where the massive object is traveling at exactly the speed of light, we see from the relativistic mass formula that its mass would be infinite, assuming its rest mass is not zero. Thus, it would take an *infinite* amount of force to push a mass up to the speed of light. Since all real forces are finite, this means that no mass-possessing object can be accelerated up to (or past) the speed of light. We again have confirmation that faster-than-light travel is not possible for human beings, since we have mass. Theoretically, we can accelerate a spaceship as close to the speed of light as we want, by progressively applying more force to

---

[13] We might at first think that Newton's second law $\boldsymbol{F} = m\boldsymbol{a}$ will still work in relativity if we let mass be the relativistic mass (γm_0) instead of rest mass (m_0). But this is not so. If the force applied is in the same or opposite direction as the mass, the resulting acceleration will obey the following equation: $\boldsymbol{F} = \gamma^3 m_0\boldsymbol{a}$. Interestingly, if the applied force is exactly perpendicular to the velocity, then the acceleration does happen to obey the relation: $\boldsymbol{F} = \gamma m_0\boldsymbol{a}$, but only in that case. In general, the formula $\boldsymbol{F} = m\boldsymbol{a}$ (unlike $\boldsymbol{p} = m\boldsymbol{v}$) cannot be salvaged merely by using relativistic mass.

it. But since the ship becomes increasingly "heavy," no finite force is sufficient to push it up to the speed of light.

What about a particle that has exactly zero rest mass? If such a particle were to travel at the speed of light, its relativistic mass from the formula above would be zero divided by zero, which is algebraically undefined. We will need a different method to calculate its relativistic mass. But from the above formula, we know this: *only a particle with zero rest mass can travel at the speed of light*. We can therefore conclude that photons (particles of light) have a rest mass of zero.

So the relativistic mass formula does not preclude the possibility of particles with zero rest mass traveling at the speed of light, but it does require that *only* particles with zero rest mass may travel at c. Nor does it automatically disallow particles to be created in a state where they are already traveling faster than the speed of light, a topic we will explore in a later chapter. But it is clear that a particle of non-zero rest mass can never be accelerated up to or past the speed of light because its mass would become infinite, requiring an infinite force.

Since $p = mv$, and since $m = \gamma m_0$, we can also express momentum in terms of the rest mass of the object as follows:

$$p = \gamma m_0 v \tag{13.5}$$

Earlier we briefly discussed the two alternative ways to define mass. In the above proof I defined mass from the momentum formula, such that $p = mv$ is true at all velocities. This has the advantage that we do not have to alter the classic formula for momentum, and it also means that mass is conserved in any isolated system. It has the disadvantage that the mass of a system is reference-frame dependent. Alternatively, many physicists choose to define mass as the rest mass. This has the advantage that mass is not reference-frame dependent (since it is always evaluated in the stationary frame of the object). But it means that the momentum formula is not $p = mv$ where m is the rest mass. But from the equation above, the momentum formula can be expressed in terms of rest mass simply by multiplying by the Lorentz factor.

Defining "mass" to be "rest mass," the formula for momentum becomes:

$$p = \gamma mv \qquad \text{(alternative notation)}$$

Many textbooks use this notation, taking "m" to mean rest mass, rather than relativistic mass, and thereby omit the zero subscript. For consistency, I will always use m_0 to refer to rest mass; any other mass is assumed to be relativistic mass, such that $p = mv$ remains true at all speeds.

Regardless of which definition of mass we take, it is clear that an object becomes more resistant to a change in velocity as its velocity approaches the speed of light. Its relativistic mass increases. But why does this occur? We know mathematically from the above proof that it must occur. But is there a mechanism? After all, the fast-moving ship still has the same number of atoms that it had when stationary. Yet, it is now "heavier" for some reason. What accounts for the extra mass?

We explore this topic in the next chapter. The answer to the question of what causes the extra mass resulted in one of the most remarkable discoveries of the twentieth century!

Chapter 14

Relativistic Energy

In the previous chapter, we proved mathematically that when an object is accelerated to high speed, it gains mass. It essentially becomes "heavier." But why does this happen? Is there an intuitive reason, or is it merely something that comes out of the math? What accounts for the fact that the high-speed ship has gained mass?

It cannot be the number of atoms; the moving ship has the same number of atoms it had when stationary. It cannot be an increase in volume, since the volume of the ship actually reduces at high speed due to length contraction. How can it be more massive? What does the moving ship have that the stationary ship does not have? The answer is kinetic energy.

The moving ship has kinetic energy; the stationary one does not. Could it be that energy itself has mass? It may seem at first that we haven't made any changes to a ship when we speed it up, and so we are surprised that it becomes more massive with speed. But really, we have added something to the ship: energy. If energy has mass, then we can easily explain why the moving ship has more mass than the stationary one; it has more energy. So let's consider the relationship between the kinetic energy possessed by the ship, and the extra mass it gains compared to its stationary value.

Derivation of the Relationship Between Mass and Energy

For this proof, we are going to limit the scenario to low velocities, much less than the speed of light. So $v \ll c$. This will allow us to use some approximations. So we imagine some object with rest mass m_0, moving at some small velocity v. Let's refer to its total relativistic mass as M, where $M = \gamma m_0$. Now since the velocity is small, the relativistic mass M will be only very slightly larger than m_0. But there is a difference. Let's call this difference m. Moreover, since the ship is in motion, it has kinetic energy. And since we are considering only velocities that are very small compared to the speed of light, we may use the classical formula as a good approximation for kinetic energy, namely:

$$E = \frac{1}{2}m_0 v^2 \tag{14.1}$$

The extra bit of mass due to this energy is m, and adds to the rest mass m_0 to give a total mass of M:

$$M = m + m_0$$

And from our formula for relativistic mass (13.4) we have:

$$M = \gamma m_0$$

Combining this with the previous, our formula for the extra mass (m) due to energy is:

$$m = \gamma m_0 - m_0$$

The Lorentz factor (5.2) can be expressed as $\gamma = \left(1 - \frac{v^2}{c^2}\right)^{-1/2}$ giving this:

$$m = m_0\left(1 - \frac{v^2}{c^2}\right)^{-1/2} - m_0$$

The middle term can be expanded into a series, using the binomial theorem. This theorem states that a quantity such as (1-x) raised to the power of -r can be expressed as follows:

$$(1-x)^{-r} = 1 + rx + \frac{1}{2}r(r+1)x^2 + \frac{1}{6}r(r+1)(r+2)x^3 + \ldots$$

If r is not an integer, the series of terms goes on forever, but as long as x is small, each term is significantly smaller than the previous term. So we can ignore all but the first few terms, and this will be a very good approximation. In our equation for the extra mass m, $x = (1-v^2/c^2)$ and r = ½. So our expression becomes:

$$m = m_0\left(1 + \frac{1}{2}\frac{v^2}{c^2} - \frac{1}{8}\frac{v^4}{c^4} + \frac{1}{16}\frac{v^6}{c^6} + \ldots\right) - m_0$$

And since we are dealing with velocities much slower than the speed of light, we can neglect higher order terms (terms in which v is raised to a high power) because they are so small. That is, since v/c is much less than 1, it follows that v^4/c^4 will be much, much, much less than one, because taking a small number to the fourth power yields a much smaller number. So this approximation will be good if we neglect all but the first two terms. So we have:

$$m = m_0\left(1 + \frac{1}{2}\frac{v^2}{c^2}\right) - m_0$$

This simplifies to the following:

$$m = \frac{1}{2}m_0 v^2 / c^2$$

But the right-hand term is simply the kinetic energy ($E = ½m_0v^2$) divided by the speed of light squared. Substituting the kinetic energy into this expression we have:

$$m = E/c^2$$

Solving this equation for E gives a remarkable result:

$$E = mc^2 \qquad (14.2)$$

Technically, we have only proved that $E = mc^2$ *for velocities much less than the speed of light.* This is because we made two approximations above.   Recall, that we assumed the classical formula for energy ($E = \frac{1}{2}m_0v^2$), which is only a good approximation when $v \ll c$. And we also neglected higher order terms in the binomial expansion of the relativistic mass formula, which is only a good approximation when $v \ll c$.

Interestingly, these two approximations exactly cancel, so that the formula $E = mc^2$ is in fact exactly true at all velocities up to and including the speed of light. The more exact proof requires calculus, and is given in Appendix A for the mathematically-adventurous reader. We have shown that the "extra" mass of a moving object due to its kinetic energy is equal to that energy divided by the speed of light squared.

That simple formula has been called the greatest discovery of the twentieth century because it has profound implications, not only for our understanding of the universe, but for technology as well. But what does it really mean?

Implications of $E = mc^2$

We have proved that kinetic energy actually has mass (relativistic mass). So when we increase the velocity of something, giving it extra kinetic energy, the mass of that object will increase a bit. If we then slow the object, removing some of its kinetic energy, the mass will return to what it was before.   The formula $E = mc^2$ means that energy is a real, physical quantity with measurable mass. It's not merely conceptual.

Moreover, this discovery was not something that Einstein merely guessed.   It was not a hunch that turned out to be right.   Rather, it was something that he mathematically proved long before it was experimentally verified, much like the proof we developed above. Merely from the definition of energy, momentum, and mass, along with the two postulates of relativity, we have proved that energy really does have mass. Or, at least, *kinetic* energy really does have mass.

What about the rest mass of an object? Could it be that the rest mass of an object is *also* due to its energy? In this case, the energy would not be kinetic (since the object is stationary), but *potential energy.* This is in fact the case (we will demonstrate this below). The total energy of an object of mass m is equal to mc^2.

Moreover, since the speed of light is a very large number in standard units, it follows that the speed of light squared is an even larger number. And since multiplying the mass by the speed of light squared gives the energy, this implies the following:

A very small amount of mass contains an enormous amount of energy.

If we could somehow harness all of the energy contained in just a small amount of mass, we could supply all of the world's energy needs. For example, the United States of America uses about 1.5×10^{19} joules of electrical energy every year. That is a huge amount of energy. What is the corresponding mass? Since $E = mc^2$, it follows that that $m = E/c^2$. So dividing 1.5×10^{19} joules by the speed of light (2.998×10^8 m/s) squared (8.988×10^{16} m^2/s^2) gives a mass of 170 kilograms. That mass corresponds to a weight of about 370 pounds. That is less than the weight of a typical motorcycle. So if we could somehow extract 100% of the energy contained in just one motorcycle due to its mass, we could power all of the electricity in the United States for an entire year!

But we are getting ahead of ourselves. The above proof only showed that $E = mc^2$ for *kinetic energy* – energy of motion. That is, the m in this formula is merely the "extra" mass that an object attains when it gains kinetic energy E. Can we also know that the *total* energy of the object (kinetic plus potential) is equal to its total mass multiplied by the speed of light squared? Yes – and there are several ways to prove this. This can be demonstrated by a mathematical proof. But it is also possible to prove it by a thought experiment as follows.

Proof That $E = mc^2$ Applies to All Forms of Energy

Photons – particles of light – have a rest mass of zero. We saw in the last chapter that this must be the case because any positive rest mass particle would require infinite force to accelerate it up to the speed of light. But photons do have momentum – this has been measured experimentally. Therefore, photons do have mass, but it is purely relativistic mass due to their kinetic energy. In other words, a photon's mass is due *entirely* to the "extra" mass that an object possesses due to its kinetic energy. If you could "stop" a photon, thereby removing all of its kinetic energy, its mass would be zero.

Conversely, electrons have a small but non-zero rest mass. So do *positrons*. A positron is a particle identical to an electron but with a positive charge. Positrons and electrons have mass even when they have no motion. If they are in motion, they will have a bit of "extra" mass due to their kinetic energy. When a positron and an electron collide, they "self-destruct" and produce a pair of photons. This has been experimentally confirmed. What happens to the energy and mass of the electron and positron when they collide and annihilate?

From the conservation of energy, the total energy before and after the collision must be the same. And from the conservation of mass, the total (relativistic) mass before and after the collision must be the same. After the collision, the total energy of the two photons E must be related to their combined relativistic mass (m) by $E = mc^2$. This is because all of their energy is kinetic – and we know this formula works for kinetic energy. But from conservation of energy, the total energy of the electron-positron pair before the collision must also be E. However, only *some* of that energy was kinetic since electrons and positrons might be moving slowly and have very little kinetic energy; the rest of their energy is potential. Likewise, from conservation of mass, the combined relativistic mass of the electron and positron must be m. Perhaps only a small part of that mass is due to the "extra" mass gained from kinetic energy; most of their mass is due to their rest mass.

Since E and m have not changed during the collision, it follows that the total energy of the electron-positron pair E, must still be equal to their combined relativistic mass multiplied by the speed of light squared. But the energy E before the collision is the total energy, which includes *both* kinetic and potential. And the total mass m before the collision is the combination of the rest-mass and the "extra" mass due to kinetic energy of each particle. Therefore, $E = mc^2$ is true of the total energy (kinetic plus potential) of the electron-positron pair when we take their total mass (rest mass plus the mass gained from kinetic energy). We can see that $E = mc^2$ applies to all forms of energy, both potential and kinetic.

So the total energy of an object or system is equal to the relativistic mass of that object or system multiplied by the speed of light squared. It follows that the rest mass of an object is caused by the sum of all its potential energy. This is a remarkable discovery because classical physics has no way to measure the absolute amount of potential energy. In classical physics, only *differences* in potential energy can be measured or calculated. Before the physics of Einstein, there was no way to know how much total energy is contained in any object. But now we know. We simply multiply its mass by the square of the speed of light.

Additional Implications of $E = mc^2$

This tells us something profound about the universe. Energy and mass are equivalent. They are basically the same thing, simply measured by different methods and using different units. The square of the speed of light may be thought of as simply a unit conversion, which enables mass to be expressed in units of energy, or energy in units of mass. Recall that this type of unit conversion was necessary when computing the spacetime interval or proper time. We had to multiply the time term by c^2 in order to combine it with the spatial terms.

Contrary to popular belief, $E = mc^2$ does *not* mean that mass can be *converted* into energy, or energy into mass. Rather, it means that mass *is* energy, and energy *is* mass. And since the speed of light is a large number, the energy contained in just a small amount of mass is enormous. Physicists have long known that different atomic nuclei have different masses. If we could harness the energy associated with the mass difference between atomic nuclei, even though the mass differences are small, this must correspond to an enormous amount of energy. Hence, atomic or nuclear power would be extremely efficient.

Consider the rest mass of a proton: 1.6726×10^{-27} kilograms. Scientists know from particle physics that if we force four protons together, two of them will convert to neutrons (and release two positrons to conserve charge) and the four particles will combine to form a helium nucleus. You might think that a helium nucleus has four times the mass of a proton. But it doesn't. It is a bit less massive. A helium nucleus has a rest mass of 6.6447×10^{-27} kilograms – not 6.6904×10^{-27} kg. So the process of converting four hydrogen nuclei into one helium nucleus results in a mass loss of 0.683 percent. The mass is carried away by radiation.

So if we could convert 1 kilogram of hydrogen (about 2.2 pounds) into helium, this would result in 0.99317 kilograms of helium. A mass of 0.00683 kilograms would escape the system in the form of radiation. How much energy would be released in this reaction? We multiply the mass by the speed of light squared to get: 614 trillion joules of energy! For comparison, a typical household uses about 40 billion joules of electrical energy in a year. If you live 80 years, your house will have used about 3 trillion joules – less than one percent the energy produced in the conversion of 1 kilogram of hydrogen to helium.   Such remarkable efficiency is why nuclear power is such an attractive option.

Kinetic and Potential Energy

It is sometimes convenient to conceptually separate the kinetic energy from the potential energy of an object or system. This is done by expanding the energy formula as follows:

$$E = mc^2 = \gamma m_0 c^2 = m_0 c^2 + (\gamma - 1)m_0 c^2$$

Here, the total energy has been divided into two components. The first component (m_0c^2) does not change with velocity. So it must be the potential energy of the object. We can also call this the "rest energy" of the object, which naturally corresponds to its rest mass.   The second component ($(\gamma\text{-}1)m_0c^2$) does depend on velocity, and goes to zero when velocity is zero. Therefore, it must be the kinetic energy (K).

$$K = (\gamma - 1)m_0 c^2$$

The total energy can then be written as the sum of the potential energy plus kinetic energy:

$$E = m_0 c^2 + K$$

This equation can also be written in terms of momentum rather than kinetic energy. The derivation follows.

First, we square the energy formula:

$$E^2 = (m_0 c^2 + K)^2 = (m_0 c^2)^2 + 2m_0 c^2 K + K^2$$

Substituting the formula for kinetic energy K we get:

$$E^2 = (m_0 c^2)^2 + 2m_0^2 c^4(\gamma - 1) + (\gamma^2 - 2\gamma + 1)m_0^2 c^4$$

This simplifies to the following:

$$E^2 = (m_0 c^2)^2 + \gamma^2 m_0^2 c^4 - m_0^2 c^4$$

We then substitute the formula for the Lorentz factor, and multiply the right-hand term by γ^2/γ^2.

$$E^2 = (m_0c^2)^2 + \frac{m_0^2c^4}{1 - \frac{v^2}{c^2}} - \frac{m_0^2c^4\left(1 - \frac{v^2}{c^2}\right)}{1 - \frac{v^2}{c^2}}$$

This makes it possible to combine the two right-hand terms and simplify:

$$E^2 = (m_0c^2)^2 + \frac{m_0^2v^2c^2}{1 - \frac{v^2}{c^2}} = (m_0c^2)^2 + \gamma^2m_0^2v^2c^2$$

However, the relativistic mass is just the rest mass multiplied by the Lorentz factor (equation 13.4): $m = \gamma m_0$. And we defined momentum to be the relativistic mass multiplied by velocity: $p = mv$. So by substitution we arrive at the following:

$$E^2 = (m_0c^2)^2 + (pc)^2 \qquad (14.3)$$

The above formula shows that the square of the total energy of an object is equal to the sum of the square of its rest energy and the square of its momentum multiplied by c. Notice the striking similarity to the Pythagorean theorem. The relationship between total energy, rest energy, and momentum (multiplied by c) is exactly the same as the relationship between the hypotenuse of a right triangle and its other sides. This provides us with an easy way to visualize the relationship between rest energy, total energy, and momentum.

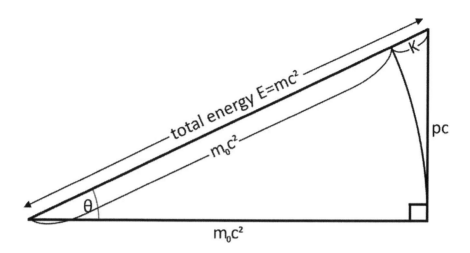

The illustration is a mathematically correct way of illustrating the relationship between the total energy, momentum, and rest energy of any object.   Given two of these quantities you can compute the third. Furthermore, it happens that the sine of the angle θ is simply the velocity of the object divided by the speed of light.

$$\sin \theta = v/c$$

The illustration also allows us to easily analyze special cases, where one of the three main quantities is zero.   For a stationary particle, its momentum (p) is zero. And so the total energy is merely the rest energy: m_0c^2. On the other hand, if a particle has no rest mass (such as a photon), then its total energy is entirely kinetic, and is given by:

$$E = pc \qquad\qquad (14.4)$$

Here, p is the momentum of the particle which, by definition, is the product of the relativistic mass multiplied by velocity. So for a particle with no rest mass:

$$E = mvc$$

But for all particles, $E = mc^2$. So we have for a particle with no rest mass:

$$mvc = mc^2$$

Diving both side by the relativistic mass (m) we find:

$$v = c$$

Earlier, we showed that any particle that moves at the speed of light must have zero rest mass. Here, we have proved that any particle with zero rest mass *must* move at the speed of light![14]

This makes sense. If we add a small amount of kinetic energy to a very massive object, its velocity does not increase very much, because mass is a resistance to a change in motion. If we add a small amount of kinetic energy to a low-mass object, its velocity can become very great. If we apply a small amount of energy to a particle with *zero* rest mass, it will jump to the maximum possible speed: the speed of light.

In the previous chapter, we saw that the formula relating rest mass to relativistic mass is not sufficient to reveal the relativistic mass of photons, because it yields zero divided by zero, which is algebraically undefined. But since $E = mc^2$, we can now compute the relativistic mass of a photon from its energy. Namely, the relativistic mass of a photon of energy E is:

$$m = \frac{E}{c^2}$$

Tachyons

We have seen that particles with positive rest mass must move slower than the speed of light. Such particles are called *bradyons*. Electrons, protons, neutrons, pions, mesons, baryons, leptons, neutrinos, quarks, and intermediate vector bosons (and almost every other particle you have ever heard of) are all bradyons because they have positive rest mass. Particles that have a rest mass of zero (and positive energy) must move at the speed of light and are called *luxons*. Of course, photons – the particles of light – are luxons. Gluons (particles that hold together an atomic nucleus) are also thought to be luxons, and so is the as-yet-undiscovered graviton.

---

[14] This is true as long as the particle has non-zero energy, and therefore non-zero relativistic mass. If the relativistic mass (m) is zero, then we may not divide both sides by m to arrive at the last step.

But consider a particle with positive energy that moves faster than the speed of light. These hypothetical particles are called *tachyons*. We know from previous chapters that such particles are not likely to actually exist in nature because faster-than-light particles can travel backward in time and potentially lead to temporal paradoxes (if the particle is capable of carrying information). But it is interesting to see what properties such particles would have if they did exist. And we can compute this based on the mathematics above.

Interestingly, tachyons would have a rest mass that is not positive, not negative, and not zero. Their rest mass would be an imaginary number: the square root of a negative number! Here is how we know this:

The total energy of any particle is given by $E = mc^2$ where m is the total (relativistic) mass. The relativistic mass is simply the rest mass multiplied by the Lorentz factor γ. So we have the following:

$$E = \gamma m_0 c^2 \qquad (14.5)$$

$$E = \frac{m_0 c^2}{\sqrt{1 - \dfrac{v^2}{c^2}}}$$

Since a particle traveling faster than light will have $v > c$, this means $v^2 > c^2$. So the term under the radical will always be negative, and thus the denominator of the right-hand side of the equation will be imaginary. Since c^2 is positive, the only way energy (E) can be positive (and not imaginary) is if m_0 is also imaginary. An imaginary number divided by an imaginary number is real. So we conclude:

$$m_0^2 < 0$$

Strangely enough, if tachyons existed, their rest mass would be imaginary. This might lead some people to instantly dismiss the possibility of such a particle. After all, how can mass have an imaginary value? And perhaps this is a legitimate objection to the existence of tachyons; they apparently don't exist and perhaps this is why. But then again, a similar argument could be used to dismiss the possibility of photons. After all, how can a real particle have a rest mass of zero? And yet photons do exist.

And if that zero-rest mass is disturbing, we would do well to remember that photons are never at rest. They must move at the speed of light (if they have any energy whatsoever). So their relativistic mass, which is the only experimentally measurable mass, is *not* zero.

Likewise, tachyons may have an absurd rest mass, but they are never at rest – by definition. We defined a tachyon as a faster-than-light particle. So hypothetically, if you could stop one, it wouldn't be a tachyon anymore. Furthermore, just like a photon, a tachyon has a *real*, non-zero relativistic mass. We know this because its energy is positive, and energy is simply relativistic mass multiplied by the speed of light squared.

We can also think of it this way. For superluminal (faster-than-light) particles, the Lorentz factor is imaginary because $v^2 > c^2$. When we multiply the imaginary rest mass by the imaginary Lorentz factor, we get the *real* relativistic mass. Moreover, the momentum is also real since it is simply relativistic mass multiplied by velocity. So providing that tachyons stay superluminal (which they must do to remain tachyons), there is no absurdity in any measurable quantity. We have real energy, real mass, and real momentum.

But what happens when we attempt to stop a tachyon? The way to stop an ordinary, slower-than-light particle is to remove its kinetic energy by applying a force. Reasoning by analogy, we might try to stop a tachyon by removing some of its energy. What happens to the tachyon? Does it slow down and become a bradyon?

Let's define E_0 as the energy of a tachyon freely moving at velocity v_0 which is, of course, greater than c. We apply a force on the tachyon to reduce its energy to a lower value, E_1. What will its new velocity (v_1) be? We have:

$$E_0 > E_1$$

Substituting the formula for energy ($E = \gamma m_0 c^2$) with the Lorentz factor expanded we have:

$$\frac{m_0 c^2}{\sqrt{1 - \frac{v_0^2}{c^2}}} > \frac{m_0 c^2}{\sqrt{1 - \frac{v_1^2}{c^2}}}$$

Each side of the inequality is a real number, because each is an imaginary number divided by another imaginary number, resulting in a real number. So we square both sides:

$$\frac{{m_0}^2 c^4}{1 - \frac{v_0^2}{c^2}} > \frac{{m_0}^2 c^4}{1 - \frac{v_1^2}{c^2}}$$

Now each side has a negative numerator divided by a negative denominator. We divide both sides by $m_0^2 c^4$. And since this is a negative number (because m_0 is imaginary), the inequality flips from > to <. (It is just as -2 > -4 but 2 < 4.) We then take the reciprocal of both sides. Of course, when we take the reciprocal of both sides of an inequality, the inequality flips *back* to the way it was (e.g.: 1/3 < 1/2 but 3 > 2). We get:

$$1 - \frac{v_0^2}{c^2} > 1 - \frac{v_1^2}{c^2}$$

We then subtract one from both sides, and multiply both sides by negative c^2. Since we are multiplying by a negative number, this again flips the inequality and we have:

$$v_0^2 < v_1^2$$

This result is very surprising, because the new velocity (v_1) is greater in magnitude than the original velocity (v_0). But we removed kinetic energy: the new energy is less than the original. Apparently, when you remove energy from a tachyon it speeds up! It behaves "backward" from an ordinary particle.

So what would happen if we removed *all* the energy of a tachyon? In this case, $E = 0$. And the only way the energy equation balances is if the denominator on the right-hand side is infinite which can only be the case if $v = \infty$. So if you could remove all the energy from a tachyon, its speed would go up to infinity, and it would be everywhere along its flight path at the same time. Strangely, though its energy would be zero, the momentum would be positive and finite, and equal in magnitude to the (imaginary) rest mass of the particle divided by the speed of light. We can see this from the energy/rest mass/momentum formula, setting $E = 0$.

So apparently, if you want to *slow down* a tachyon, you must *add* energy to it. Simply repeat the above algebra, but reverse the inequality so that the new energy is greater than the original: $E_0 < E_1$. All subsequent steps have a reversed inequality, but the algebra is otherwise unchanged. This results in the strange conclusion that the new velocity will be *less* than the previous. Adding kinetic energy to a tachyon slows it down. But how much slower?

As we add energy to the tachyon, its speed drops, but it remains superluminal. This must be the case because the energy E is positive; in fact, it is becoming quite large as we continue to add energy. And the only way to have positive energy with a particle of imaginary rest mass m_0 is if the Lorentz factor is also imaginary, so that the two imaginary numbers cancel. Hence, no matter how much energy we add, the velocity continues to drop, but $v > c$. As the added energy approaches infinity, the Lorentz factor gets larger in magnitude, reaching infinity in the limit as v is reduced to c. In other words, it would take an infinite amount of energy to slow a tachyon down to the speed of light!

Just as you cannot accelerate a bradyon up to the speed of light because it would require infinite energy, you likewise cannot slow down a tachyon to the speed of light because this also would require infinite energy. The speed of light is therefore the unattainable upper speed limit for a bradyon, the unattainable lower speed limit for a tachyon, and the required speed for a luxon.

The tachyon does slow down as we add energy, but that extra energy adds to the relativistic mass. The slower a tachyon travels, the "heavier" it becomes, making it harder to slow it further. Sound familiar? This is just what happens when we try to accelerate a particle of finite rest mass up to the speed of light. Its relativistic mass increase makes further acceleration more difficult.

Well, tachyons probably do not actually exist in nature. But it is interesting that the mathematics of relativity continues to work perfectly well with superluminal particles. The results are astonishingly self-consistent, though somewhat surprising and counterintuitive.

Relativistic Mass and Gravity

In the previous chapter, I mentioned that most physicists prefer to define mass as "rest mass" and avoid using the concept of relativistic mass, preferring instead to redefine the momentum formula. But when we consider gravity, we find that relativistic mass is a very useful concept.

In classical physics, there are actually two different definitions of mass: inertial mass, and gravitational mass. So far, we have only discussed inertial mass. This type of mass is defined in terms of resistance to a change of momentum under a force. In the classical limit, $F = ma$. So mass is the constant of proportionality between force and acceleration.   But mass can also be defined in terms of the force of gravity. Newton's law of gravity states:

$$F = \frac{-GMm}{r^2}$$

Here, F is the force of gravity, M is the gravitational mass of one object, m is the gravitational mass of the other object, r is the distance between them, and G is the "universal gravitational constant." This gravitational constant is a number that sets the strength of gravity for the entire universe and which has a value of 6.67×10^{-11} Nkg^{-2}m^2.

So gravitational mass is defined as an object's ability to produce or respond to a gravitational force. However, inertial mass is defined as an object's resistance to a change in motion. Seldom do physicists concern themselves with the difference, because all experiments conducted to date have not been able to find any measurable difference between these two types of mass. As far as we know, the gravitational mass of any object is always equal to its inertial mass.

For this reason, most physics textbooks don't distinguish between the two. We simply refer to an object's mass, without any need to specify whether this is inertial mass or gravitational mass since they are always apparently the same. But why these two different definitions of mass should always be equal is the source of some debate.[15]

How does this topic relate to relativistic mass? Those physicists who prefer not to use the concept of relativistic mass would say that an object only seems to be more resistant to a change in motion at high speed because the momentum formula is not $p = mv$, but rather $p = \gamma mv$. The apparent strangeness, they argue, is in the momentum, not in the mass.

---

[15] When Einstein began incorporating gravity into relativistic physics, he partially answered this question. By redefining gravity, not as a force, but as a change in the geometry of spacetime, Einstein showed that an object's response to gravity must be proportional to its inertial mass. But this still doesn't fully explain why an object's ability to produce gravity (to alter the metric of spacetime) ought to be proportional to its inertial mass.

They would say that the high-speed particle simply *behaves as if* it has more inertial mass, when it really does not.

But what about gravitational mass? As we increase the speed of a particle, does its ability to produce and respond to gravity also increase? If so, then it seems that we have every reason to believe that this extra mass is *real*, and not just an artifact of using the wrong momentum formula. After all, we would not expect the momentum formula to affect gravity. So we shall now attempt to answer the question of whether or not the "extra" (apparent) mass due to an object's kinetic energy produces a gravitational field.

Consider photons – particles of light. We saw above that these particles have a rest mass of zero. So any mass they have is entirely relativistic mass due to their kinetic energy. We all agree that photons have momentum; this has been experimentally measured. Those who do define mass as rest mass would say that photons merely have momentum because they have energy, but they have no mass. On the other hand, if relativistic mass is real, then it should produce gravity. So the question is: do photons produce a gravitational force on other objects?

Photons do respond to gravity; they "fall." We don't notice this much in our earthly experience because light is so fast and Earth's gravity is so weak that the effect isn't noticeable. However, when light passes by a massive object like the sun, its path is bent noticeably by the gravity of the sun. Astronomers have observationally confirmed this by measuring the apparent positions of stars during a solar eclipse.

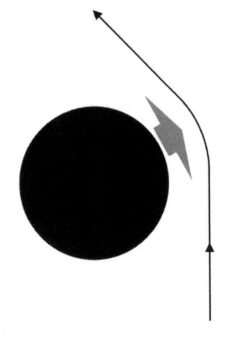

So imagine photons traveling in the positive *y*-direction. They have some momentum **p** which is entirely in the positive *y*-direction as well. They pass very close to the sun, and their path is deflected. Their velocity is now partially in the negative x-direction. Likewise, their momentum has changed direction slightly. The x-component of the momentum vector is now negative. And the y-component of their momentum is slightly less than it was before they were redirected.

In our thought experiment, we imagine that the sun and the photons are the only things in the universe. Thus, they constitute an isolated system. But from conservation of momentum, the total momentum (both magnitude and direction) of an isolated system cannot change with time. And yet it seems we have a violation here. The initial momentum of the system was entirely in the y-direction. But after the photons deflect, their momentum is now partially in the negative x-direction. How do we resolve the paradox?

The key is to remember that the *total momentum of the system* cannot change with time. Thus, if the photons (which constitute only part of the system) gain momentum in the negative x-direction, it follows that the sun must gain the same magnitude of momentum in the positive x-direction. If the photons lose a bit of momentum in the y-direction, then the sun must gain the same bit of momentum in the y-direction.

So the only way that momentum will be conserved is if the sun begins to move slightly in the positive y- and positive x-direction, as shown by the grey arrow. In other words, the sun is pulled by the gravity of the photons! This way the total x-component of the momentum of the system remains zero, and the total y-component of the system remains whatever it was to begin with. In reality, the acceleration of the sun due to the gravity of light will be far too small to measure. But our thought experiment proves that the effect must exist in order for momentum to be conserved.

So we conclude that photons, although they have no rest mass, do have mass because they produce a gravitational influence on other bodies (though the effect is but slight). This seems to confirm that relativistic mass is quite real – it is both inertial and gravitational.

This does not conclusively disprove the alternative view, because its supporters might claim that the formula for gravity needs to be adjusted for high speed, much as they say the formula for momentum is. Really, it is simply a matter of definitions. In any case, the extra mass due to a particle's kinetic energy behaves like mass in all respects – both in terms of inertia and gravity. Treating relativistic mass as real allows us to use the momentum formula without modification, and explains why photons both respond to and produce a gravitational field.

Chapter 15

Introduction to General Relativity

When Einstein published his initial discoveries in the field of relativity in 1905, he did not yet know how to handle the effects of gravity. He spent the next ten years working on general relativity – an adaptation that includes the effects of gravity. This may seem like a very small modification to the theory. After all, it seems that we are merely adding one force to the theory. Yet, the addition of gravity changes the entire game. Unlike special relativity, general relativity requires a level of mathematical rigor that is not easy to grasp.

For this reason, we cannot delve into general relativity with the same depth as special relativity. Nonetheless, we can discuss at least qualitatively how gravity interacts with space, time, and mass. This will allow us to explore some fascinating phenomena, such as black holes and gravitational time dilation. We will, in some special cases, even be able to discover the mathematical formula for a few of these effects.

The Equivalence Principle

We saw in the last chapter that mass can either be defined as gravitational mass (the ability to produce and respond to the force of gravity), or inertial mass (a resistance to a change in motion under a force). Remarkably, these two different types of mass are always identical in all known cases. This astonishing coincidence bothered Einstein. Why should two fundamentally different definitions of mass always give exactly the same answer? There must be a connection between the two.

The equality between inertial mass and gravitational mass is why physicists normally do not bother to specify which type of mass they are discussing. We simply refer to an object's mass, since gravitational mass and inertial mass are always equal. One result of this equality is that all objects must necessarily fall at the same rate (same acceleration) under gravity in the absence of other forces (such as air resistance). We proved this by a *Gedankenexperiment* in an earlier chapter – showing that anvils and hammers must fall at the same rate. We can also prove it mathematically, but only if gravitational mass and inertial mass are the same:

Let m be the mass of an object that we let fall to the ground. The force of gravity on the object is given by Newton's law of gravity:

$$F = \frac{-GMm}{r^2} \tag{15.1}$$

Here, M is the mass of the Earth (5.972×10^{24} kg), r is the distance to Earth's center (roughly 6.37×10^6 meters), and G is the universal gravitational constant (6.67408×10^{-11} m^{3}kg^{-1}s^{-1}). The negative sign indicates that the force of gravity is attractive; it is opposite in direction to the distance vector from the source of gravity. Under the classical limit, the object will experience acceleration toward Earth according to Newton's second law:

$$F = ma \tag{15.2}$$

If gravity is the only force on the falling object, then we may substitute this equation into the first equation:

$$ma = \frac{-GMm}{r^2}$$

Dividing both sides by m gives the acceleration of the falling object:

$$a = \frac{-GM}{r^2} \tag{15.3}$$

Which for the surface of Earth is 9.8 m/s² in the downward direction.

The wonderful thing about the result (15.3) is that the mass of the falling object (m) does not appear in the final formula. This means that the rate at which something falls is independent of its mass. Hence, all objects fall at the same rate in the absence of air or other forces. But this proof assumes that gravitational mass (used in the formula for gravity) is the same as inertial mass (used in Newton's second law). If this were not the case, then in the last algebraic step above, dividing both sides by inertial mass, the *m* terms would not cancel.

It is a remarkable result. An object of large mass *requires* a greater force to accelerate it to a certain speed than a less massive object. Yet, the object of large mass *receives* a greater force from Earth's gravity – an amount that is precisely what it needs to accelerate it at the same rate as the lesser mass. Why is this? This nagging coincidence led Einstein to believe that gravity is unique and quite different from the other fundamental forces. Could it be the case that gravity is not really a force?

Consider a rocket ship in deep space, far away from any sources of gravity. The rocket fires its engine downward and begins moving upward due to the reaction force. The rocket continues to move upward with progressively greater speed, as its engines continue to fire. Suppose a passenger on the rocket (who is stationary relative to the rocket) holds out two objects side by side – a very massive object and a much "lighter" one, at time T_1 (left frame in the figure). What happens when he lets go? Those two masses would continue to drift at constant velocity – whatever velocity the rocket had at the moment the passenger let go. They would be initially stationary relative to the rocket. But the rocket is accelerating. Its upward speed is increasing, and so the floor of the rocket eventually moves at a greater speed than the two objects and will approach them (at T_2). It will therefore eventually catch up and collide with the two objects at T_3.

But what would the passenger in the rocket see? He is accelerating with the rocket. So from his point of view, the two objects began to move downward toward the floor the moment he let go of them. They started moving slowly at first, but then began moving progressively faster until they hit the floor. Although there is no gravity in deep space, nonetheless, the passenger would see the two objects "fall" to the floor as if being pulled by gravity. What is even more wonderful is that both would hit the floor at the same time! The rate at which the two objects fell did not depend on their mass. But this is exactly what happens with gravity! Unless you look out a window, you cannot tell merely from dropping objects whether you are stationary on the surface of a planet with gravity, or on an accelerating rocket far away from gravity.

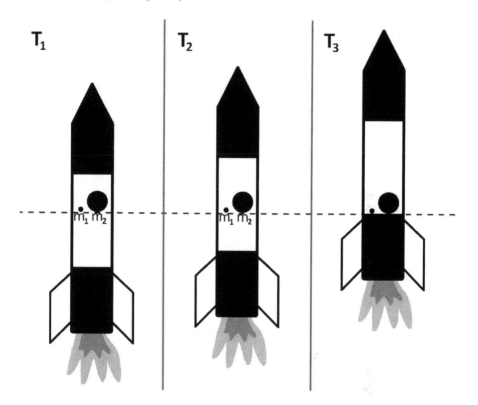

T_1 T_2 T_3

It seems that gravity mimics the effect of acceleration. A stationary frame under the influence of gravity is just like an accelerating frame that is not under the influence of gravity. Indeed, any experiment you can do on the surface of the Earth involving gravity, you can do on a rocket ship that is accelerating upward at 9.8 m/s^2. The effects of gravity and acceleration appear to be equivalent in terms of how the laws of physics work. This effect, which Einstein discovered, we call "the equivalence principle."

The principle also works if we consider a rocket that is not accelerating, but merely floating at constant speed through space. Under the equivalence principle, this situation will appear identical to one in which a rocket is freely falling toward Earth under the acceleration of gravity. In both rockets, if a passenger releases two masses, they will remain stationary relative to him – they appear to float rather than fall. For the rocket in deep space this is obvious; there are no forces on the objects, so they will not accelerate. But for the freely falling rocket, the force of gravity induces the same acceleration on the masses as it does the ship and all passengers. They all fall at the same rate, and therefore remain stationary relative to each other. Apparently, the laws of physics work the same for a constant-speed object in deep space as they do for an object in free fall under gravity. This reasoning led Einstein to a remarkable conclusion:

A free fall frame is inertial.

In special relativity, an inertial frame is a non-accelerating, non-rotating frame. An inertial frame, therefore, moves at constant velocity or is stationary. But special relativity neglected the influence of gravity. Gravity, Einstein realized, essentially "redefines" what constitutes an inertial frame.

Imagine that Eric stands on the surface of Earth while his sister Linda skydives from an airplane. During the first few seconds after jumping (before wind resistance becomes significant), Linda is in free-fall. She is accelerating, moving progressively faster toward the surface of Earth. Which observer is inertial?   Under special relativity, Eric would be considered inertial since his reference frame is non-accelerating, and Linda would be non-inertial since she is accelerating. General relativity reverses these roles. Linda is inertial because she is moving freely with no external forces on her (apart from gravity – which is not truly a "force" by Einstein's definition). But Eric is not inertial because he is not in freefall. He does have an external force pushing on him; the surface of the Earth pushes upward on him, preventing him from falling toward Earth's center.

Recall that the laws of physics work identically in any inertial frame. This is the first postulate of relativity. The second postulate is that the speed of light is constant with respect to any inertial observer. General relativity adds a third postulate: a free-fall frame is locally inertial.

By "locally inertial" we mean that the frame is inertial at sufficiently small scales. The approximation breaks down at large scales. For example, suppose the rocket ship were about as wide as the diameter of Earth. In that case, it would be possible to distinguish whether the rocket is stationary on the surface of Earth or accelerating in deep space. An object placed on the left side of the stationary rocket's floor would pull toward the right, since it will be gravitationally attracted to Earth's center, whereas an object on the right side would pull toward the left for the same reason. Furthermore, an object placed at the top of the rocket would be significantly farther from Earth's center, and so it would fall at a slower rate than objects near the floor. These are "tidal effects" and would not be experienced by the rocket accelerating in deep space. But if the rocket ship is sufficiently small, such tidal effects would be negligible, and it would be impossible to tell from experiments inside the ship whether it is stationary on Earth or accelerating in deep space.

Gravitational Time Dilation

One wonderful result of the equivalence principle is that gravity causes time dilation. Recall from the chapter on the Doppler effect, that an accelerating frame will experience time dilation along the direction of acceleration. Namely, clocks near the bottom of an upward-accelerating rocket will tick slower than clocks near the top. And from the equivalence principle, we understand that a stationary object under the influence of gravity will behave just like an accelerating object that is *not* under the influence of gravity.

Consider a rocket that is stationary on the surface of Earth, with passengers near the bottom at point A and passengers near the top at point B. The distance between A and B is h. In chapter 11, we found that if this rocket were accelerating in deep space with acceleration a, then the passengers at A would experience time dilation relative to passengers at B by this formula (11.6):

$$\frac{\Delta T_B}{\Delta T_A} = 1 + \frac{ah}{c^2}$$

That means, the same effect must occur for the stationary rocket on the surface of Earth. It behaves as if it were an accelerating rocket, with an acceleration equal to the free-fall acceleration under Earth's gravity. Namely, $a = g$, where $g = 9.8$ m/s^2 at sea level (15.3). Therefore, if we have two clocks on the surface of Earth, and one is at a higher elevation than the other by h, then the clock that is lower will be time-dilated relative to the higher clock by:

$$\boxed{\frac{\Delta T_B}{\Delta T_A} = 1 + \frac{gh}{c^2}}$$

(15.4)

As a result of gravitational time dilation, a clock at sea level will tick a bit slower than a clock at high elevation. The effect will be small, of course, and will require atomic clocks to measure the difference in time dilation. But this can be done; and it has been done. The National Institute for Standards and Technology (NIST) has an atomic clock in Boulder, Colorado, at an elevation of 1660 meters above sea level. There is also an atomic clock at the United States Naval Observatory in Washington, D.C., with an elevation of about 80 meters. From our formula, the ratio of time dilation between the two clocks is $1+1.7 \times 10^{-13}$. So every twenty-four hours the clock at NIST gains about 15 nanoseconds (15 billionths of a second) relative to the clock in Washington, D.C. This has been experimentally confirmed.

The fact that gravitational time dilation has been experimentally tested and confirmed gives us great confidence in the physics of Einstein. Apparently, our logical reasoning and mathematical computations in the previous chapters have been correct. Do note, however, that this formula for the difference in time dilation due to gravity is an approximation. It is based on the time dilation due to acceleration that we derived in the Doppler effect chapter; and we used the approximation that acceleration (g) and height (h) are relatively small, such that $gh << c^2$. This is called the "weak field approximation."

Furthermore, the above approximation gives only the relative ratio of time dilation between two points at different distances from the Earth's center. But what is the total amount of time dilation for an observer on Earth's surface compared to a stationary observer in deep space that is far away from any source of gravity? The person in deep space experiences no time dilation because there is no gravity, and we suppose that both observers are stationary relative to each other. So there is no time dilation due to velocity. The only effect will be gravitational time dilation. Consider the following thought experiment.

Imagine that we are far from the Earth, and we drop something so that it falls to Earth. In the limit as our distance to Earth approaches infinity, the velocity of the falling object at every point along its journey will be the "escape velocity" at that distance from Earth. We normally think of escape velocity as the outward directed velocity necessary to overcome the force of gravity and thereby leave the system without falling back. But time is symmetric in this situation, so the escape velocity can also be taken to be the *inward* directed velocity of an object dropped from infinity. From classical physics, we know that the formula for escape velocity is:

$$v_E = \sqrt{\frac{2GM}{r}} \qquad (15.5)$$

Here, G is the universal gravitational constant, M is the mass of the Earth, and r is the distance from Earth's center. Since the distance r is in the denominator, the escape velocity diminishes with increasing distance from Earth. But at the Earth's surface (r = 6371 km), the escape velocity is 11,186 meters per second, which is 25,022 miles per hour.

So an object dropped from an infinite distance would have an increasing velocity as it falls toward Earth, given by the above formula for any given value of r. But this is a classical physics formula. We might wonder what modifications will be necessary to make this formula correct for relativity. What does the formula look like when we include the effects of time dilation and length contraction? But remember the third postulate of relativity: a free fall frame is locally inertial. This object is in free fall. Therefore, it is locally inertial, and there will be no strange effects of relativity. So this classical formula is exactly right in relativity, and needs no modification.

Instead, it is the *stationary* observer standing on Earth's surface that will experience time dilation relative to the observer in deep space. The stationary observer is not in a free-fall frame, but moves at escape velocity relative to the object that was dropped from infinity. So the time dilation he experiences relative to the observer in deep space will be the same *as if he were moving relative to that observer at escape velocity.* We therefore use the escape velocity to compute the Lorentz factor from equation 5.2 and equation 15.5:

$$\gamma = \frac{1}{\sqrt{1 - \dfrac{v_E^2}{c^2}}}$$

$$\gamma = \frac{1}{\sqrt{1 - \dfrac{2GM}{rc^2}}}$$

Thus, the time dilation of an observer stationary on Earth will be such that his clock ticks ΔT_A when the equivalent clock in space ticks ΔT_B where $\Delta T_B = \Delta T_A \gamma$. By substitution we arrive at our answer:

$$\frac{\Delta T_B}{\Delta T_A} = \frac{1}{\sqrt{1 - \dfrac{2GM}{rc^2}}} \qquad (15.6)$$

This formula for gravitational time dilation is more exact than the approximation we found previously, since we did not require that $gh << c^2$. The formula applies for any spherical source of gravity of mass M.

How much gravitational time dilation do we experience on Earth relative to someone in deep space? Substituting the appropriate numerical values into the formula, we find that clocks on Earth tick slow by 69 billionths of a percent. Over the course of 24 hours, this adds up to a difference of 0.00006 seconds. So you will live very slightly longer on Earth than you would in deep space due to time dilation. How much longer? if you live to be eighty years old, you will live an extra 1.76 seconds due to time dilation under Earth's gravity.

Unlike velocity time dilation, gravitational time dilation is non-symmetric. Recall that when two inertial objects are in motion relative to each other (far away from any source of gravity), each observer will find that the other person's clock is slower than his or her own. It is counterintuitive but true. But with gravitational time dilation, the observer in deep space will find that clocks on Earth's surface tick slower than his clock, whereas observers on Earth's surface will find that clocks in deep space tick *faster* than their own. The reason for the lack of symmetry is that the observer on Earth's surface is *not* in an inertial reference frame, but the observer in deep space is.

Gravitational time dilation is always a function of *gravitational potential* which we can think of as a sloped depression surrounding all massive objects. Imagine a bowling ball placed gently in the middle of a trampoline. The indentation it makes represents gravitational potential. The dent is deepest right at the position of the bowling ball, and diminishes with distance. The concept of *potential* is closely related to the concept of *potential energy*. The energy necessary to pull something away from a source of gravity is the gravitational potential *energy* that the object gains during the journey. The gravitational potential is simply the gravitational potential energy divided by the mass of the object that is moved.

So imagine carrying a bowling ball up a hill. That will take some energy. Imagine carrying a feather up the hill. That will take less energy because the feather is light. But the difference in gravitational potential between the bottom and top of the hill is fixed. It is the same for the bowling ball and the feather because when we divide their potential energy gain by their mass, each will give the same answer. In relativity, the gravitational potential surrounding any mass is always negative because it takes energy to pull something away from that mass due to gravity.

Every point in space has a gravitational potential associated with it, which is determined by the gravity produced by the mass of all surrounding objects. The difference in time dilation between any two points is related to the difference in gravitational potential between those two points. So the closer you are to a large mass, the slower time will flow.

Gravitational Redshift

Since time dilation occurs in a gravitational field, there will be a shift in the frequency of light between an observer on the surface of Earth and a stationary observer in deep space. The frequency of light is the number of wave peaks *per second* that pass by some reference frame. However, the person on Earth's surface has longer seconds than a person in deep space due to time dilation. Therefore, an observer in space will see fewer peaks per second by his clock than an observer on Earth's surface will measure by his clock. A lower frequency means a longer wavelength. And so the observer in deep space will see any light from Earth's surface as slightly redshifted.

The situation here is exactly analogous to the *transverse* Doppler shift we covered in an earlier chapter. This is because the distance between the person on Earth and the stationary observer in deep space does not change. Consequently, any change in the frequency of the light waves must be due entirely to time dilation; there is no change in the light travel time. Therefore, we may use the transverse Doppler formula with the Lorentz factor we found for gravitational time dilation:

$$\frac{\lambda_O}{\lambda_S} = \frac{1}{\sqrt{1 - \dfrac{2GM}{rc^2}}} \tag{15.7}$$

The effect on Earth is miniscule. But it is more substantial on objects of greater mass, particularly on massive objects that also have a small radius, so M/r is large. This is the case with white dwarf stars. These have a mass comparable to the sun, but a radius comparable to the Earth. Gravitational redshift has been observed on the white dwarf star Sirius B. The observed value is in agreement with the predictions of general relativity.

Both the gravitational time dilation formula and the gravitational redshift formula we derived above assume that both observers are stationary relative to each other and to the Earth. If we consider that one of the observers is in motion, things become more complicated. In that case, the effects of velocity time dilation will add to the effects of gravitational time dilation. And the Doppler effect will further alter the wavelength of light which is also experiencing a gravitational redshift. So in real world scenarios, time dilation and redshift are almost always a combination of velocity effects and gravitational effects. With typical stars and galaxies, the Doppler shift is generally so much larger than the gravitational redshift, that we may neglect the latter.

Redefining Geometry

Simply by the recognition that a free-fall reference frame is locally inertial, we were able to derive the above equations. A more detailed investigation of the effects of general relativity would involve some fairly intensive mathematics in order to deal with a type of geometry that we do not ordinarily experience. But we can at least qualitatively describe what happens when we add the effects of gravity to relativity.

Recall from an earlier chapter, we introduced the spacetime interval, which is a bit like a "distance" through the four dimensions of space and time. We can think of the dimensions of space and time like legs of a right triangle, and the spacetime interval is the hypotenuse. The one difference is that the square of the time dimension is subtracted from the spatial terms, rather than added to them. General relativity redefines the spacetime interval to include the effect of gravity. This effect changes the geometry of spacetime so that the spacetime interval is no longer conceptually a straight line.

The spacetime interval is analogous to the Pythagorean Theorem (other than the fact that the time dimension adds in reverse). But the Pythagorean Theorem only works for right triangles in a plane. These triangles are "well-behaved" and match our intuition. Their internal angles always add up to 180 degrees. But consider another type of triangle, one that is painted on a sphere.

Imagine a right triangle on a sphere, where the three sides are arcs along the sphere. If the triangle is very small, it will behave much as a triangle in a plane. The internal angles will add to approximately 180 degrees, and the sides will deviate only slightly from the Pythagorean Theorem. But a larger triangle would deviate substantially.

Now imagine a triangle on a sphere, where point A is on the equator, point B is also on the equator but east of point A by one-fourth the circumference of the sphere. Point C will be the north pole, as in the figure. The angle made by arc AC and arc AB will be a right angle. But so will the angle between arc AC and CB and the angle between AB and CB. This right triangle has *three* right angles, and so its internal angles add to 270 degrees, not 180 degrees. Furthermore, its hypotenuse (which could be defined as any one of the three arcs) will not equal the sum of the square of the other two legs, but will instead be equal to each of the other two legs. This is a right triangle *and* an equilateral triangle – something that is not possible in a plane.

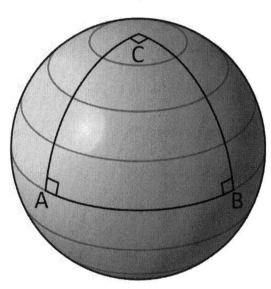

Imagine an ant walking due north from point A, while another ant walks at the same speed due north from point B. These ants are on initially parallel paths. And parallel lines never intersect. Right? Yet these ants will meet at the north pole at a 90-degree angle. Was this because some external force pushed on them, altering their direction? No. It happened because of the geometry of the sphere.

Likewise, in general relativity, gravity is not a force that pushes or pulls objects in space. Rather, it is a description of the local four-dimensional geometry of spacetime. Mass causes spacetime to be curved sort of like a sphere, rather than flat like a sheet of paper. Two masses with identical velocities that are on initially-parallel paths will eventually intersect, not because there is some force of gravity between them, but rather because their mass causes the geometry of spacetime to be slightly curved.

For this reason, the Pythagorean Theorem no longer defines the hypotenuse of a right triangle. The shortest distance between two points is no longer a straight line. However, these are both good approximations in cases where the effects of gravity are small, just as a very small triangle on a sphere very nearly obeys the Pythagorean Theorem. General relativity is all about understanding the geometry of spacetime. It can be summarized as follows: mass curves spacetime, and spacetime tells mass how to move.

This geometry is very difficult to visualize for two reasons. First, space is curved in four dimensions. But typically, we can only visualize three dimensions. The number of dimensions is the number of lines you can make intersect at one point and all be at right angles to all other such lines. No matter what you do, you can only picture three such lines. But the universe has four. Second, the time dimension is mathematically reversed from the three spatial dimensions. And it is difficult for us to understand or visualize how this works geometrically.

Einstein was able to describe the relationship between mass and the curvature of spacetime by a set of equations that are called the "Einstein field equations." There are ten field equations, but they can be written as a single formula using Einstein summation notion. The constituents of the combined formula represent a 4x4 matrix, for a total of sixteen equations; but six of these are redundant. Thus, only components on and above the matrix diagonal need to be computed. The Einstein field equations allow us to compute the differential form of the spacetime interval (ds) in the presence of a local gravitational field. They are very difficult to solve in all but the most basic scenarios. And so we will not go into any further detail. But we will address some of the results of these equations.

General Relativity of the Universe

One particularly interesting aspect of general relativity is that it can describe the average curvature of spacetime for the entire universe, given the mass distribution in the universe. Of course, we don't know precisely what the mass distribution of the universe is. Technology allows us to map the positions of the brighter galaxies. But there may be additional matter in the universe that remains unseen. As an approximation, astrophysicists often assume that the mass in the universe is isotropic (the same in all directions) and homogeneous (roughly the same density everywhere), at least on the largest scales. This allows them to solve the Einstein field equations on a cosmic scale, at least as an approximation.

Alexander Friedmann was the first to solve Einstein's field equations correctly for the universe under the assumption of an isotropic, homogenous mass distribution. He made this discovery in the early 1920s. In 1927, George Lemaitre independently arrived at the solution. In the 1930s, Howard Robertson and Arthur Walker confirmed this solution, and proved that it was the only possible solution under the given assumptions. For this reason, the geometry describing the curvature of spacetime for a universe with mass evenly distributed is called the Robertson-Walker metric. Sometimes all four names are included: the Friedmann-Lemaitre-Robertson-Walker metric.

The solution implies that the universe cannot be eternally static; it must either expand or contract. This was contrary to the mainstream scientific consensus of the early 1920s. Most astronomers at the time believed that the distant universe was static, as most people had believed for thousands of years. Even Einstein himself was initially reticent to accept Friedmann's solution, but later conceded that it was correct.

So is our universe expanding or is it contracting? Observations by astronomer Edwin Hubble in the late 1920s found that the light from most galaxies is redshifted. Moreover, the farther a galaxy is found from our position, the more redshifted it is. The redshift implies that the galaxy is receding from ours. Since the recession rate is proportional to distance, this implies that all galaxies see others receding from them too – a universal expansion. Friedmann was right – our universe is expanding. Perhaps Christians should have anticipated this result in light of many Scriptural passages that seem to teach that the heavens are being stretched out, e.g. Isaiah 40:22, 42:5, 44:24, 45:12, 51:13; Job 9:8, 37:18; Psalm 104:2; Jeremiah 10:12, 51:15; Zechariah 12:1. In any case, it seems that the Bible anticipated the discoveries of Hubble, Friedmann, and others by several thousand years.

The geometry of spacetime is determined by the rate of expansion and the average density. The density of the universe is the mass in a particular region of space divided by the volume of that region. The density can vary tremendously from place to place. But on the largest scale, the density is thought to be basically uniform – the assumption of homogeneity. Friedmann found that the geometry of the universe would fall into one of three categories depending on the cosmic density and expansion rate. But he did not know what the average density of the universe is, nor the expansion rate. So he published all three solutions, leaving the density as an unknown parameter. The three solutions are called *open*, *closed*, and *flat*.

If the density of the universe is high compared to its rate of expansion, then the closed solution applies. In this case, the geometry of spacetime is a bit like a sphere. So if we were to construct a cosmically-large triangle, we would measure the internal angles as adding to *more* than 180 degrees – just like a triangle painted on a sphere. In a closed universe, initially parallel lines eventually intersect, just like the longitude lines on a sphere intersect at the poles, though they are parallel at the equator. Imagine a giant spherical region of our universe containing many galaxies. If the galaxies are equally distributed, you might expect that doubling the radius of the sphere should result in *eight times* as many galaxies in the new larger sphere since volume goes as r^3. But in a closed universe, the new sphere will have somewhat *less* than this.

In addition to geometrical effects, the density to expansion rate ratio also affects the dynamics of the universe. In a closed universe, the gravitational attraction between all the galaxies is "larger" than the outward rate of expansion. The result is that gravity will eventually halt the expansion, and the universe will begin to collapse. A closed universe *cannot* exist forever.

Conversely, if the density of the universe is low compared to its rate of expansion, then the universe is said to be "open." The geometry of spacetime can be compared to a saddle. If you draw a triangle on a saddle and measure its internal angles, you will find that they add to *less* than 180 degrees. So a cosmically-large triangle in an open universe will likewise have less than 180 degrees. In an open universe, initially-parallel lines will diverge. A cosmic sphere with twice the radius of another one will contain *more* than eight times the number of galaxies. In an open universe the expansion rate is larger than the gravity of all the galaxies; so the universe will continue to expand forever.

Exactly in between these two cases is the "flat" universe. In this case, the density of the universe exactly matches the expansion rate. The universe expands at exactly its own escape velocity. Geometry behaves "normally" in a flat universe; all triangles have exactly 180 degrees, and a giant sphere that is twice the radius of another one will contain eight times as many galaxies. Parallel lines neither intersect nor diverge, but remain exactly parallel. In a flat universe the expansion rate exactly matches the combined gravity of the galaxies, so that gravity will constantly slow the expansion, but never quite stop it.

In physics, density is usually represented by the lowercase Greek letter rho (ρ). For a given expansion rate, the cosmic density necessary to make the universe flat is called the critical density (ρ_c). The ratio of the actual average density of the universe (ρ) to the critical density is called the density parameter, and is symbolized by a capital omega (Ω). So by definition, $\Omega = \rho/\rho_c$. For a flat universe, $\Omega = 1$. For a closed universe, $\Omega > 1$. And for an open universe $\Omega < 1$.

The Real Universe, Dark Matter, and Dark Energy

So which solution applies to our universe? To find the answer we need to measure the expansion rate of our universe, and then measure the average density. Astronomers have been able to measure the expansion rate of the universe with decent accuracy by comparing the distances to galaxies with the corresponding redshift. The expansion rate is approximately 70 (km/s)/Mpc. The units are "kilometers per second per mega parsec." A parsec is 1.917×10^{13} miles. This rate means that if two galaxies were separated by 1 Mega parsec (1 million parsecs), then the galaxies would be moving away from each other by 70 kilometers per second due to the expansion of the universe. This expansion rate is called the Hubble constant and is designated by H_0. The zero subscript indicates that this is the expansion rate at the present time; recall that gravity is constantly slowing the rate of expansion over time.

The average density of the universe is much more difficult to estimate. We can estimate the number of stars in a galaxy, and we can measure the gas and dust which also contribute to a galaxy's mass. We then multiply this by the number of galaxies in a volume of space. If so, we would find that $\Omega = 0.04$. This is much less than one, so the universe is open, it would seem, at first glance. But the answer may not be quite so simple.

Not everything in space glows, or reflects light, or gives evidence of its existence by blocking the light of other objects. Some mass might be undetectable by telescopes; dark, compact objects or massive particles that do not interact strongly with matter: "dark matter." If dark matter exists, such undetectable mass will contribute to the cosmic density, and invalidate our previous estimate of Ω. But how can we possibly detect mass that is practically invisible? Even invisible mass produces a gravitational effect on other visible mass. Therefore, by measuring the motion of visible objects, we can compute the pull of gravity on them, and deduce the total mass of the system.

When we apply such computations to the motions of stars in galaxies, and the motions of galaxies in clusters, we find that $\Omega = 0.27$. So the universe is still apparently "open," yet the density is higher than we would have suspected by counting only visible objects. What is remarkable is how much higher the new estimate is! If the number is to be believed, then there is actually far more dark matter in the universe than visible matter, about six times more. It is humbling to think that with all we know about astronomy, we have not yet even detected that which makes up most of the universe.

But things become even stranger when we consider geometric effects. Recall that the curvature of spacetime causes geometry to deviate from what we normally expect. The internal angles of triangles do not necessarily add to 180 degrees, and doubling the radius of a sphere does not necessarily result in capturing eight times more volume. Therefore, it is also possible to estimate Ω by observing how geometry works on cosmic scales.

When we use geometric techniques to estimate the density parameter (Ω), we find that it is very nearly 1, in which case the universe is flat. But how can this be, since the matter in the universe, both visible and dark, contributes only 0.27 to Ω? Where is the remaining 0.73 coming from?

It could be that spacetime itself plays a role. If spacetime exerts a sort of "pressure," it could accelerate cosmic expansion, and produce the very geometric effects that we observe. This property of spacetime is called the cosmological constant, and is denoted by the Greek uppercase lambda (Λ). If Λ has any value other than zero, it will curve spacetime, much the way matter does.[16] But what mechanism could possibly cause such a gravitational repulsion?

One possibility that physicists have proposed is that spacetime itself has energy. Since $E = mc^2$, this energy will necessarily have mass.

---

[16] Interestingly, Einstein first introduced the cosmological constant in order to preserve the old idea of a static universe. After physicists discovered that Einstein's equations implied that the universe must either expand or contract, Einstein invoked the cosmological constant as a "fix" to counter gravity and prevent cosmic collapse. But after redshifts confirmed the universe is indeed expanding, Einstein dropped the cosmological constant, calling it "the biggest blunder" of his life. Now, observations suggest that the cosmological constant may exist after all – not to halt collapse, but to accelerate expansion.

Mass produces a gravitational effect, curving spacetime. The idea that empty space has non-zero energy is called "dark energy." Since dark energy (if it exists) is a property of spacetime, it is the same everywhere. Thus, it will produce no net gravitational attraction on visible bodies. This explains why it cannot be detected by observing orbital motions, the way dark matter can.

But Einstein's equations indicate that dark energy would exhibit a type of outward pressure, which would actually tend to accelerate the expansion of the universe rather than halting it. Whether expansion of the universe slows or accelerates will depend on the exact ratio of dark energy to "ordinary" mass. The details are beyond the scope of this text. But it is interesting that most astronomers now believe that the universe is essentially flat, and that ordinary matter comprises only about 4% of the density of the universe. Whether this turns out to be correct, only time will tell.

A "Big Bang"?

So far, all that we have discussed is good science. We have moved from the well-established principles (such as the basic tenets of general relativity), to reasonable inferences (such as dark matter), to the more tentative (such as dark energy). This is science because it is testable and repeatable in the present. As we devise new experiments and observations, it is likely that we will eventually either confirm or refute ideas like dark energy. But some people have suggested that general relativity proves, or at least implies, that the universe started in a "big bang," 13.8 billion years ago. This is not true, nor is it scientific. But it deserves some discussion.

In 1927, when Lemaitre proposed the idea that the universe was expanding, he assumed that it had been created with size at some time in the past, and that it was now a bit larger due to such expansion. This is probably correct, and is what most creation scientists believe today. But Lemaitre had a philosophy of *methodological naturalism*. This is the belief that scientists should ignore God in any scientific discussion: that they should do research as if they were atheists, regardless of whether or not God actually exists. Such a philosophy is contrary to Scripture of course. The Bible indicates that God constantly upholds His creation (Hebrews 1:3). Science is possible only because God has chosen to uphold the universe in a consistent and law-like fashion.

But Lemaitre, though he believed in God, began contemplating the evolution of the universe from a standpoint of naturalism. If God had not supernaturally created the universe in the past, then we might be tempted to extrapolate the universe's present rate of expansion back to a time prior to creation. This was exactly what Lemaitre reasoned. He supposed that the expansion of the universe might be extrapolated back to a time when the universe had essentially no size at all! In 1931 he published this idea, that the universe had sprung from a point, many millions of years ago. This was a preliminary version of a model we now call "the big bang."

First, we note that Lemaitre's supposition is not observational science. It is not something that can be tested or observed in the present. We cannot repeat the origin of the universe, or directly see if it was ever compressed into a point. Such suppositions lie beyond the scope of the scientific method.

Second, it is important to note that the big bang is not something that we would conclude from the science of general relativity *without the additional assumption of naturalism*. There is nothing in Einstein's field equations that requires the universe to have started from a point or to be billions of years old. And nothing in general relativity is inconsistent with the biblical history that God supernaturally created the universe a few thousand years ago.

So it would be a mistake to claim that general relativity proves or implies a big bang. It doesn't. It is the *assumption of naturalism* coupled with the (correct) observation that the universe is expanding that leads some people to assume that the universe started as a point. But there is no scientific reason to believe this. And there are very good scientific reasons to doubt the big bang. These go beyond the scope of this book. But for further reading, please see BiblicalScienceInstitute.com.

Conclusions

Einstein's field equations are difficult to solve in most scenarios. Nonetheless, there are some situations we can examine using the equivalence principle, such as gravitational time dilation and redshift, without explicitly solving the Einstein field equations. We have already seen that we can estimate gravitational time dilation and redshift just from the equivalence principle. And we have discussed how general relativity can be applied as an approximation to the entire universe.

But there is another particularly interesting scenario that we can discuss, one that we can analyze using the equivalence principle. This is the subject of the next chapter: a black hole.

Chapter 16

Black Holes

In the previous chapter, we found that an object that is dropped to the Earth from far away will fall to Earth at the escape velocity. We saw that the formula for escape velocity (15.5) is:

$$v_E = \sqrt{\frac{2GM}{r}}$$

Since the distance from Earth's center (r) is in the denominator, larger values of r imply smaller values of v_E. So escape velocity is greatest at the surface of Earth, and gets smaller as we go out to greater distances. An object dropped to the Earth from far away will fall in such a way that at every point throughout its journey, it is traveling at the escape velocity for that distance from Earth. If we could instantly reverse its direction, the falling body would coast away from Earth, ever slowing but never stopping. The escape velocity represents the speed at which an object's kinetic energy is exactly equal to the negative of its gravitational potential energy.

Consider a very massive source of gravity, like the sun. The mass of the sun is 1.989x10^{30} kilograms, and its radius is 6.963x10^8 meters. So at the surface, the escape velocity is 617 km/s, or 1.38 million miles per hour. That's pretty fast. But if we could somehow compress the sun down to a much smaller size, the escape velocity at its surface would be much greater. At some small radius, the escape velocity would be equal to the speed of light. To compute this radius, we solve the above equation for r

(where we use $r = R_S$ to represent this "special" radius) in the case where $v_E = c$. The algebra is straightforward, giving this result:

$$R_S = \frac{2GM}{c^2} \tag{16.1}$$

This term (R_S), representing the distance from the center of a massive object at which the escape velocity is equal to the speed of light, is called the "Schwarzschild radius." The term is named after German physicist Karl Schwarzschild (in English this is normally pronounced as "shwawrts-shild"), who first derived the above solution to Einstein's field equations.[17] For the mass of the sun, the Schwarzschild radius is 1.8 miles. If we could compress the sun to an even smaller radius than 1.8 miles, then its surface escape velocity would be *faster* than the speed of light. What does this mean?

At the very least, it means we would not be able to see the sun. An object must travel at or faster than escape velocity in order to escape from the gravity of a mass. But the escape velocity for this compressed sun is *greater* than the speed of light. And light cannot travel faster than the speed of light. Hence, light cannot escape the compressed sun. Any light emitted by the sun's surface would immediately fall back to the sun's surface. It would never reach our eyes. Therefore, the sun would become invisible. It would be a "dark star."

Second, if the sun were compressed to less than 1.8 miles in radius, *nothing* could ever leave its surface. The logic is straightforward. If an object must travel faster than light to escape, and if nothing can travel faster than light, then nothing can escape. Any journey to visit the surface of such a dark star would be a one-way trip.

Interestingly, the concept of a dark star actually preceded the physics of Einstein by over a century. In 1783 John Michell proposed the idea of a star that is so massive that its own light could not escape. He supposed that any light emitted would fall back to the surface, thereby rendering the star invisible. Michell further proposed that some of these dark stars might be part of a binary star system, in which case we could detect them by their gravitational influence on their visible companion. Essentially, he suggested that we should look for bright stars that appear to orbit nothing.

---

[17] This solution is valid for any non-rotating, electrically neutral mass. The resulting geometric description of the spacetime surrounding such a mass is referred to as the *Schwarzschild metric*.

Michell's ideas were brilliant and insightful, particularly in light of the fact that the physics necessary to understand such "dark stars" had not yet been discovered. Yet, Michell was right. There are objects whose mass is sufficient to prevent light from escaping. Today, we refer to these as "black holes." Indeed, astronomers have discovered a number of black holes by the very method that Michell proposed; they are part of a binary star system and gravitationally influence the path of their visible companion star. But there are some differences between Michell's idea of a dark star, and our modern understanding of a black hole. General relativity allows us to describe what would actually happen to any object near the Schwarzschild radius.

The River Model of Black Holes

To understand what effect gravity has on space, the analogy of a lake can be very helpful. The surface of the water of the lake represents empty space in the reference frame of some inertial observer. Boats on the lake move relative to the water at some velocity. We could imagine that some speedboats have their throttle stuck at a certain speed "c" and their rudder also is stuck so that these boats always move in straight lines at constant speed. These represent photons. They cannot travel at any other speed than c, and they always travel in straight lines. All other boats must travel slower than c, but they can get as close to c as we wish. So far, this scenario is analogous to what we would find in special relativity.

To understand how gravity works, imagine that there is a hole at the bottom of the center of the lake, and water is allowed to drain through this hole into a lower chamber. We imagine that there is no "swirl" or rotation; the water simply speeds directly toward the hole. Water very near the edge of the lake would be virtually unaffected by the hole. But water near the hole would move quite rapidly closer to the hole, like a river. The hole represents a massive body. Its gravity does not pull on the boats directly; rather it pulls on the water so that the water anywhere in the lake is progressively pulled toward the center. The boats simply go along for the ride. As the water nears the hole, its speed progressively increases, just like a body falling toward a massive object.

What would happen to the "photon speedboats" if they were to pass relatively close to the hole? Their path would deflect. They move at speed c relative to the water; but the water near the center of the lake is moving quite rapidly toward the hole. The change in direction of water flow as the photon boats pass by would cause a change in their direction, even though their rudder does not change. The closer the photon speedboats come to the center of the lake, the more their path will be deflected from a straight line.

If the hole were large enough, the water near the hole would drain at a very rapid rate. At a certain point very near the hole, we imagine that the water might even reach a speed of c, which represents the speed of light. The distance from the hole where the water reaches speed c represents the Schwarzschild radius (R_S). Inside the Schwarzschild radius, the water would be falling toward the hole even faster than c. This doesn't violate any rule, because the speed limit of c applies to the speed that *boats* are allowed to travel *relative to the water*. The speed at which the water may flow is unrestricted.

Photons and the Schwarzschild Radius

What would happen if we placed a photon boat very near the hole, and aimed it directly away from the hole? The boat moves at speed c relative to the water away from the hole, but the water moves at high speed toward the hole. If the boat were farther away from the hole than the Schwarzschild radius, then the (outward) speed of the boat will be greater than the inward speed of the water. Like salmon swimming upstream, the boat will slowly drift away from the hole, gaining speed as it gains distance. Likewise, when outward-directed light is emitted just outside the Schwarzschild radius of a black hole, it slowly drifts away from the black hole, gaining speed as it moves away.

Already some astute readers might think, "But didn't we establish earlier that the speed of light is always constant? How can it move so slowly?" The key is to remember that the speed of light is constant *relative to an inertial observer*. And in general relativity this is further restricted to a *locally* inertial observer. An observer who is stationary relative to the black hole will indeed see the light moving slower than c – but a stationary observer is *not inertial* in general relativity. Instead, the free-fall observer is inertial. And such an observer would be falling into the black hole at the escape velocity for the given distance. From *his* point of view, the photon moves at exactly speed c.

Conversely, suppose we placed the boat closer to the hole than the Schwarzschild radius, then the inward speed of the water will be *faster* than the outward speed of the boat relative to the water. In this case, the photon boat – even though it is directed outward – will move inward toward the hole. This represents outward directed light that is emitted *inside* the Schwarzschild radius. Such light will be pulled "backward" and fall deeper into the black hole.

A particularly interesting case occurs when we place the outward-directed photon boat exactly at the Schwarzschild radius. In this case, the outward speed of the boat exactly matches the inward speed of the water. So the photon boat will be stationary! This is exactly what happens to outward-directed light that is emitted at the Schwarzschild radius of a black hole. The light remains forever stationary, neither approaching nor receding from the black hole.

Matter is always slower than the speed of light. Therefore, any matter at or inside the Schwarzschild radius must be pulled further into the black hole. It cannot move faster than speed *c* relative to the water, and the water is moving faster than speed *c* toward the hole. Thus, whatever events happen within the Schwarzschild radius can never affect anything outside that radius. The Schwarzschild radius thus represents a type of *horizon*. (A horizon is something that you cannot see beyond.) Events inside this horizon cannot have any causal influence on events outside the horizon. For this reason, the Schwarzschild radius of a black hole defines a spherical surface called the "event horizon." The event horizon is not a physical surface, but a conceptual one. It is a spherical region that divides the inside of the black hole from the outside.

Inside a Black Hole

The event horizon represents the "point of no return" for a black hole. Anything outside the event horizon has the possibility of escape. Inside the Schwarzschild radius, inertial space is falling toward the center of the black hole at a speed greater than the speed of light (relative to a hypothetical observer who is stationary relative to the mass of the black hole). And since no mass can move through space faster than light, any material object will necessarily be pulled closer to the center of the hole.

This necessarily includes the surface of the dark star itself. For any star whose surface is inside its event horizon to remain the same size, the material would have to travel outward through the infalling space faster than the speed of light, which of course it cannot do. Therefore, once a star has been compressed so that it fits entirely inside its own event horizon, it *must* continue to collapse all the way down to a point. It does not matter what the star is made of, or how rigid or resistant to compression the material is. To remain stationary inside a black hole is the general relativity equivalent of accelerating a mass past the speed of light. It cannot be done.

The mass responsible for the black hole must collapse all the way to a point of zero size in the exact middle of the event horizon. This central point is called a "singularity." Such a singularity must form inside a black hole because any alternative would violate the second and third postulates of general relativity. Anything that falls within the event horizon of a black hole is destined to intersect the singularity and become part of it, adding slightly to its mass.

The term "black hole" generally refers to the volume of space surrounding the singularity within the event horizon. To fall past the event horizon is to fall into the black hole. But all the mass of a black hole is in its singularity. Since the singularity has positive mass, and a volume of zero, its density can be considered infinite.[18]

Unfortunately, you can never see a singularity. The singularity of a black hole is always inside its own event horizon. Therefore, no light or any other information-carrying signal can be transmitted from a singularity to the outside world so that we may learn about it. Black hole singularities are always "clothed" by an event horizon. The alternative would be called a "naked singularity" – one that does not have an event horizon surrounding it, and may therefore be seen. The *cosmic censorship hypothesis* claims that naked singularities cannot exist in the universe. This hypothesis is widely accepted, but has not yet been proved.

Outside a Black Hole

If we were near a black hole, but well outside of the event horizon, what would we see? Our first guess might be "nothing." After all, light cannot escape a black hole, and so it is necessarily invisible. But much of the rest of the universe is not. A black hole would block and absorb the light from any stars that are behind it from an observer's point of view. And so we

---

[18] General relativity does not include the effects of quantum mechanics, the latter of which describes how the universe behaves at very small scales. Quantum mechanics disallows a perfectly defined position with known momentum; hence the singularity may actually have a non-zero (though sub-atomic) volume. Unfortunately, quantum mechanics cannot describe what happens in an intense gravitational field, the very kind associated with a singularity. So there is some mystery here, and some questions which are not yet answered. Modern physics therefore cannot fully describe what happens in a singularity, though it can describe what happens just outside one.

would see an *absence* of stars or of any light whatsoever in a small circular region of the background universe directly behind the black hole. Ironically, the black hole is revealed by what we *don't* see.

Furthermore, recall that the bending of spacetime around a black hole will deflect the path of light, just as the "photon speedboats" are deflected by the velocity of the water near the hole in a lake. The closer a photon passes to the event horizon, the more severely its path will be bent. Therefore, when we look toward the black hole, background stars will appear distorted. They will seem to have shifted away from the black circle, an effect that becomes very severe for stars that appear near the limb. In fact, some stars will produce multiple images because some of their light will be bent around the left side of the black hole and deflected to the right to reach the observer, whereas other light will be bent around the right limb and be deflected to the left to reach the same observer. The effect will also concentrate the light somewhat so that at least one of the star's images will appear brighter than it would if there were no black hole at all. Black holes act like a distorted magnifying glass with a black spot in the middle.

If a bright star happened to be directly behind a black hole with respect to some observer, the light would be bent equally around the black hole at the same angle in all directions. The observer would see the star not as a single object, but as a complete and very bright circle surrounding the black hole. In honor of Einstein, this effect is 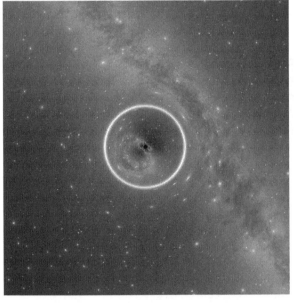 called an "Einstein ring." The effect is simulated in the figure – a star directly behind the black hole forms a bright ring. The Milky Way galaxy in the background is a bit off center, and therefore forms two images: the large image to the right outside the ring, and a small curved image just to the lower left of the black hole inside the Einstein ring. Stars that are far in angle from the disk of the black hole appear nearly undistorted.

The bending of light by gravity is called "gravitational lensing." It is a real phenomenon, and does not necessarily require a black hole; any sufficiently massive object will do. Galaxies, for example, contain hundreds of billions of stars. The combined gravity of all that mass is able to bend the light from background galaxies quite significantly. We now have many pictures of galaxies that have been "split" into multiple images as their light passes close to a foreground galaxy on its way to Earth. In some cases, we get an Einstein ring, or sometimes an "arc" – a section of an Einstein ring.

Classical Orbits

What would happen to the orbit of Earth if we could somehow compress the sun into a black hole? The answer is... nothing. Earth would continue to orbit exactly as before, and so would all the planets. Their orbital velocity is determined by the mass of the sun, which has not changed. As long as we are far away from the Schwarzschild radius, orbits are well approximated by classical physics. There are no strange and wonderful effects of relativity so far away.

But as we get close to a very massive object, the effects of relativity begin to emerge. The planet Mercury is close enough to the sun that the effects of general relativity are actually detectable, though they are *very* small. One of these effects is called *precession*. To understand this effect, we need some background information on how orbits work in classical physics.

In classical (Newtonian) physics, orbits between two bodies are always exact ellipses.[19] An ellipse is a "squashed" circle. It is defined as the set of all points in a plane that have an equal sum of the distance from two specified points. Let me illustrate this. Imagine two points on a rectangular plane: A and B. These two points are called "foci" (plural). Each one is a "focus." Now imagine a third point some distance from them. This third point will be some distance from A, and some distance from B. These two distances will add to a particular value. An ellipse is the set of all points whose distance from A and distance from B sum to the same value. Moving the foci farther apart but keeping the total distance the same results in a very squashed ellipse. The degree of "squashedness" is called *eccentricity*,

---

[19] This explanation is simplified. Technically, an exact ellipse is only the case for two point masses. Large masses that rotate or are non-spherical can cause the orbit to deviate slightly from a perfect ellipse.

and can never exceed a value of 1 for a closed orbit. Moving the two foci closer together reduces eccentricity. A circle is a special case of an ellipse where the two foci have been moved together, so that they are one point. A circle has an eccentricity of zero.

It turns out that the orbits of the planets are all ellipses, where the sun is at one focus.[20] Generally, planets have very low eccentricities, so that their orbits are nearly circular. The eccentricity of Earth's orbit is only 0.0167 – pretty close to a circle. Since the orbits are ellipses, they have a point of closest approach to the sun, which we call the *perihelion*. Likewise, they have a point of maximum distance, which we call *aphelion*. The orbits of most planets are so nearly circular that it would be hard to detect where perihelion and aphelion are simply by eye. Mercury is the exception. It has the highest eccentricity of the eight classical planets, with a value of 0.2056. This is high enough that, when Mercury's orbit is plotted, the human eye can detect that Mercury is noticeably closer to the sun at perihelion than at aphelion.

Orbits involving only two bodies are closed, which means the perihelion and aphelion always occur at the same two points in space. But our solar system has more than two bodies. The planets mutually pull on each other, which causes the perihelion and aphelion of orbits to drift slightly. If perihelion occurs at an angle of zero degrees on one orbit, it might be 0.1 degrees on the next orbit, then 0.2, 0.3, 0.4, and so on. The ellipses gradually rotate around the sun. The effect is tiny, but it does exist. Classical physics correctly predicts the rate of precession for all the planets – except Mercury.

Relativistic Orbits

Mercury is close enough to the mass of the sun that the effects of general relativity cannot be neglected. In general relativity, elliptical orbits naturally precess, even if only two bodies are involved. This is due to the curvature of spacetime around the massive body. This effect is small, but it adds to the precession due to the gravitational interaction with the other planets. Thus, Mercury's precession is different from what it would be if general relativity were ignored.

In fact, Mercury's orbit precesses by 1.556 degrees per century. But Newtonian physics predicts that Mercury should precess at only 1.544

---

[20] The other focus is empty. Technically, the focus is the center of mass of the solar system. But since the sun dominates the mass of the solar system, we will not concern ourselves with the small difference.

degrees per century. The difference of 43 arc-seconds is due to the effect of general relativity. Relativity correctly predicts the total precession of Mercury, and this has been confirmed observationally.

What happens if we consider orbits that are fairly close to a black hole? The effect of precession becomes severe. An object in an elliptical orbit sufficiently close to a black hole might precess by 30 degrees or more on each orbit! Over time, the path would resemble a Spirograph-type pattern. But far away from the black hole, precession becomes negligible.

Orbits in which the closest approach lies at least three times farther than the Schwarzschild radius are *stable*. A stable orbit is one in which a very small thrust in any direction simply results in a new slightly modified orbit. Therefore, all orbits that come no closer than $3R_S$ are "safe." They might precess by an enormous amount. But a small thruster malfunction will merely result in a slightly changed orbit; it will not send the object into the black hole or into deep space.

But that changes once we pass closer to the star than $3R_S$. If our spaceship were to pass inside this distance, we must be very careful not to thrust in the wrong direction. A slight bump could send us into the black hole, or hurtling away into deep space, depending on the direction of the bump. Orbits inside $3R_S$ are unstable.

A very interesting situation occurs at $2R_S$. In general relativity, the effect of precession increases as we get closer to the black hole. At $2R_S$ the rate of precession exactly matches the angular orbital speed at perihelion. So what does this mean? Imagine a very elliptical orbit around a black hole. Suppose the aphelion is very far, and the perihelion is exactly twice the Schwarzschild radius. As the object falls toward perihelion, the precession rate of the ellipse increases. When the object reaches perihelion, the precession rate of the ellipse matches the orbital speed of the object. So the object remains at perihelion as it orbits the black hole. It would appear from the outside that the object is orbiting in a perfect circle at $2R_S$. A very small thrust backward will send the object plummeting into the black hole. But, a very small thrust forward will break the pairing, and send the object back out toward aphelion. An orbit at $2R_S$ is therefore called a *marginally bound orbit*.

This is a fascinating situation and a very convenient one. It turns out that an orbit at $2R_S$ has a total energy of zero; the object's kinetic energy is the negative of its gravitational potential energy. This means that we could drop into this orbit without using any fuel. Such an orbit is not possible in classical physics. When our spaceship is very far from a black hole, an infinitesimal thrust in the correct direction will allow us to fall into a circular orbit at $2R_S$, where we could study the black hole as long as we want, without using any fuel. When we're done, a small thrust forward will send our ship back out to a safe distance. Of course, this energy-free orbit is inside the $3R_S$ limit, which means it is unstable. A very small thrust in the wrong direction would send our ship into the black hole.

Another particularly interesting orbit occurs at $1.5R_S$. At this distance, light can orbit the black hole! If aimed in just the right direction, light will orbit in a perfect circle at $1.5R_S$. The fact that light can orbit a black hole has interesting implications for what we would see if we were very near the event horizon. Light that is reflected or emitted by our space ship can potentially circle the black hole and travel back to the ship, where it might pass through a window and into our eyes. Hence, by looking out a window, we could see our own ship some distance away! Depending on the orbital parameters, we might even see multiple images of our ship at various distances and directions.

Inside $1.5R_S$ orbits are not possible, stable or otherwise. If your ship passes closer to the black hole than this distance, you need to activate your engines immediately. Otherwise your ship will pass into the event horizon. And if that happens, you need to make sure that you are right with God, because you are about to meet Him.

Dropping an Object into a Black Hole

Suppose our spaceship is hovering over a black hole at some safe distance. Our engines are thrusting, keeping us stationary relative to the black hole. What would happen if we released a small object and let it fall into the black hole? What would we see? Let's suppose the object is a brightly glowing clock, synchronized with our clocks on the ship.

As the clock falls closer to the back hole, it would appear to tick slower. This is an optical effect because the light from each subsequent tick has to travel a greater distance to reach our ship than the light from the previous tick did. Furthermore, light emitted near the event horizon is traveling "upstream" through the infalling inertial space, so it travels more slowly from our perspective than the speed of light in deep space. The clock also becomes very reddish in color, due to the redshift of the light as it escapes the gravitational well. The clock also becomes quite dim. This is due to the optical time dilation; if the clock emits one billion photons per second in its own frame, but its seconds are stretched to an hour in our frame, then we see only 278 thousand photons per second. The number of photons per second that we detect gets progressively fewer, until we struggle to see anything at all. The object would dim and fade from our sight.

Suppose we had the technology to compensate for the redshift and the progressive dimming of the clock – special goggles that brighten the object and "unredshift" it. This way, we can continue to watch the clock until it passes through the event horizon. Right? But that moment never comes. As the clock gets closer to the event horizon, its seems to fall more *slowly*. Its recession gets slower, and slower, never quite reaching the event horizon. It apparently never falls through. What is going on here?

In reality, the clock fell through the event horizon right on schedule. But (outward-directed) photons emitted by the clock just before it fell through continue to struggle to escape against the infalling inertial space long after this. It is just like the "photon boats" trying to climb up the water falling into the hole in the middle of the lake. Recall, the speed of inertial space falling into the black hole at the event horizon is the speed of light. So just outside that radius, space is falling inward at *almost* the speed of light. The photons emitted there might take hours just to move one inch away from the black hole. As they move farther away, the speed of infalling space is less, and they make greater progress. Eventually these photons reach our eyes in the spaceship above, and we see an image of the clock as it was just before it fell in, even though it really fell in many hours earlier. Of course, photons emitted even closer to the event horizon might take years, or even centuries to reach the ship. So in principle, we could *always* see some photons coming from the clock!

By the same logic, *in principle* you can see an image just above the event horizon of *everything* that has ever fallen into a black hole. This is because there will always be a few photons from the object, as it was just before it fell into the black hole, streaming into space. Of course, these images become increasingly redshifted and increasingly faint with time. So in practice, this effect doesn't last very long. In a short amount of time, the image becomes far too faint and too redshifted to be seen. But one consequence of this effect is that *you can never actually watch something cross the event horizon*. Instead, it will merely redshift and fade from view.

Rotating and Charged Black Holes

When we considered the scenario of collapsing the sun to within a radius of 1.8 miles, we did not address the fact that the sun rotates. The above scenario describes what would happen for a non-rotating black hole. But since the sun rotates, if we compressed it into a much smaller volume, its rotational speed would increase tremendously. This is due to a principle of nature called the conservation of angular momentum. We can see this effect demonstrated when an ice skater pulls her arms in while spinning; her rate of spin increases.

The amount by which the spin increases is related to the difference between the initial size and the final size. Compressing the sun from its current radius of 432,700 miles down to 1.8 miles would result in an enormous spin-up. The spin rate of the surface would be a substantial fraction of the speed of light, even if escaping material carried off some of the angular momentum. Therefore, relativistic effects due to rotation cannot be ignored. The solution to Einstein's equations for a *rotating* compact mass is called the Kerr metric. It is similar in many respects to the Schwarzschild metric of a non-rotating black hole. The same sorts of effects of time dilation, redshifts, and precessing orbits all exist. But there are some differences.

Since Kerr black holes have been generated by compressing a rotating mass, they are not spherically symmetric like a Schwarzschild black hole. Instead, they have an axis of rotation and a corresponding equator. In general relativity, a massive, rapidly rotating object will "twist" the spacetime surrounding it – an effect known as *frame dragging*. We might have expected something like this from our lake analogy. Water seldom travels in a perfectly straight path down the drain; it usually swirls a bit, particularly right near the drain. This is roughly analogous to what a Kerr black hole does to the inertial space surrounding it.

The strength of frame dragging depends on the angular momentum of the black hole. Theoretical calculations have shown that there is a maximum possible rotation for a black hole, at which the frame dragging effect is maximized. A particle dropped straight toward the equator of a Kerr black hole will not fall straight in, but will spiral in, gaining some tangential velocity in the same direction that the Kerr black hole rotates. Objects thrown toward a Kerr black hole opposite its rotation will actually reverse direction before falling in.

This changes some of the orbital parameters from what they were in the Schwarzschild case. Photons no longer orbit at $1.5R_S$. They can actually orbit a bit closer or farther depending on the direction in which they orbit, and the strength of frame dragging. Likewise, the region of marginally-bound or marginally-stable orbits will depend not only on distance but also on direction. A ship can safely orbit closer to a Kerr black hole by going in the same direction as the frame dragging.

Frame dragging causes another difference between the Kerr black hole and the Schwarzschild black hole. For a non-rotating black hole, a spaceship can (theoretically) remain stationary at any distance above the event horizon by exerting sufficient thrust. But for a rotating black hole, a ship very near the event horizon will be carried along by frame dragging. There is a minimum distance from a Kerr black hole at which a ship can remain stationary relative to the distant stars by exerting thrust. This is called the "static limit." It lies at a greater distance from the singularity than the event horizon around the equator of a Kerr black hole, and coincides with the event horizon at the poles. The region in between the static limit and the event horizon is called the *ergosphere*.[21] No object can remain stationary in the ergosphere. If you enter this zone, you will inevitably be dragged along in the direction of rotation.

A few other interesting effects distinguish a Kerr black hole from its Schwarzschild counterpart; these concern what happens around the singularity. In a Kerr black hole, the singularity is thought to be a ring, rather than a point. Theoretical calculations show that spacetime interior to this ring is unstable, suggesting that the Kerr metric doesn't apply in this volume. Moreover, the Kerr metric implies that a second inner event horizon forms, which is called the Cauchy horizon. However, this too is unstable.

Most objects in the universe have some non-zero angular momentum. Thus, if they were to collapse into a black hole, it would surely be a Kerr black hole rather than a Schwarzschild black hole. Technically, a Schwarzschild black hole is a special case of a Kerr black hole where the spin happens to be zero. Astronomers expect that black holes in the real universe are of the Kerr variety, with a spin in between zero and the maximum possible limit.

A third type of black hole deals with what happens when the mass has a strong electric charge. In this scenario, spacetime is described by the Reissner-Nordström metric. When charged particles are brought together, there is positive potential energy between them, which increases without limit as the distance goes to zero. And since $E = mc^2$, this potential binding energy naturally manifests as mass, which changes the metric somewhat from the Schwarzschild solution. It causes a second horizon to form internal to the black hole, a Cauchy horizon. This is similar to what happens in the Kerr metric, although the charged black hole remains spherically

---

[21] Ergosphere means "energy sphere." It is so named because it is theoretically possible to extract some of the black hole's rotational energy from the ergosphere.

symmetric. Black holes with substantial charge are very unlikely to exist in nature, because (1) most of the universe is electrically neutral, and (2) they would preferentially attract the opposite charge and quickly become neutral.

A fourth solution considers a black hole with rotation *and* electrical charge, essentially combining the Kerr and Reissner-Nordström metrics. This is the Kerr-Newman solution. General relativity predicts that this metric can describe any black hole that can possibly exist, because black holes can have only three externally observable independent properties: mass, angular momentum, and charge.[22] We can think of the previous three metrics as special cases when charge is zero (Kerr), when angular momentum is zero (Reissner-Nordström), or when both are zero (Schwarzschild). But again, the only black holes likely to exist in the real universe are of the Kerr variety.

Black Holes in the Real Universe

The above discussions have been primarily hypothetical. We have investigated what physics indicates would happen if we could compress material down to the Schwarzschild radius. But does this actually happen? Do we have observational evidence that black holes actually exist? In fact, the evidence is quite abundant. Black holes come primarily in three different mass categories: stellar mass black holes, intermediate-mass black holes, and supermassive black holes.

Stellar mass black holes are so named because their mass is comparable to that of a star. Astronomers have discovered a number of stellar mass black holes by the gravitational tug they induce on their companion star – the very method that John Michell proposed two centuries ago. This also allows us to estimate the mass of the invisible object. All such black holes are more massive than the sun, with a typical value between 3 and 20 solar masses.

One such black hole is associated with the x-ray source Cygnus X-1. At this location, a blue supergiant star and a black hole orbit around their common center of mass every 5.6 days. Based on the orbital data, the invisible companion must have a mass of about 15 solar masses, which is consistent with other stellar mass black holes.

---

[22] This is postulated in the "no hair" theorem, so named from its succinct summary: black holes have no hair.

Most astronomers believe that stellar mass black holes form when a massive star explodes – a supernova. Such an explosion ejects the outer layers of the star into space, but compresses the inner layers to the Schwarzschild radius, forming a black hole. Only very massive stars are able to do this. If the core of a star that remains after a supernova is less than 3 solar masses, no black hole will form. This is because the outward pressure of the material (neutrons) is sufficient to prevent the surface from falling below its own Schwarzschild radius. No further collapse is possible at that point. But for stellar cores significantly greater than 3 solar masses, the pressure from neutrons is insufficient to overcome the acceleration of gravity. And so the star collapses without limit and forms a black hole.

On the other end of the scale are supermassive black holes. Their mass ranges anywhere from hundreds of thousands to over a billion times that of the sun. Astronomers find these in the center of a galaxy, and it may be that every galaxy has one. Our own galaxy, the Milky Way, unquestionably has such a black hole at its core. This is indicated by the motions of stars near the core. From their orbits, we know that the mass of the black hole is approximately 4 million solar masses. How such a black hole could form is the subject of much debate. It seems likely that the universe was created with such black holes already in them, perhaps to act as gravitational anchors for stars near the core.

Less well-established is a class of intermediate-mass black holes, with a typical mass between 100 and 1 million solar masses. These may exist in the disks of galaxies and at the cores of some globular clusters. Only a handful have been detected. And many such claims are disputed.

Falling into a Black Hole

What would it be like to explore a black hole? Suppose we had a ship with powerful engines capable of hovering just above the event horizon of a black hole. What would we experience? Suppose that we decided to enter the black hole. What would that look like? First, we need to decide what mass category of black hole we want to visit. And this brings us to the topic of tidal forces.

Many people have some experience with tides from seeing the ocean advance and recede over the course of time at a beach. The tides are caused partially by the gravity of the sun, but mainly by the gravity of the moon. Recall from Newton's formula for gravity, the force of gravity is inversely proportional to the square of the distance to the source. So the closer you are to the moon, the greater the force of gravity.

This means that the portions of Earth's oceans directly below the moon experience a greater pull of gravity than Earth's core, since they are slightly closer to the moon than Earth's core. Therefore, they will be pulled toward the moon a bit – high tides. We also see high tides on the opposite side of the Earth – but why? This portion of Earth's ocean is farther from the moon than Earth's core; therefore, Earth's core is pulled more strongly by the moon, and is pulled "away" from the ocean.

Meanwhile, the portions of Earth's ocean that are at a 90° angle relative to the moon will be pulled the same amount as Earth's core, but not quite in the same direction. They will be pulled at a small angle relative to Earth's center, as shown in the left panel of the figure. When we subtract the net force of the moon's gravity, we are left with the forces seen in the right panel. Namely, the Earth is stretched vertically and compressed horizontally. These are tidal forces.

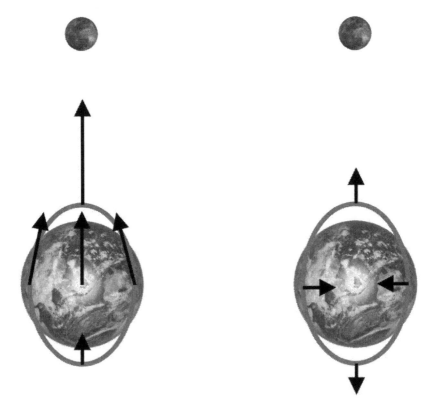

Likewise, a person orbiting a black hole will experience tidal forces. He will be stretched vertically, and compressed horizontally. The same effect occurs for a person standing on Earth: it simply is far too small to be noticed. But technically, the force of gravity on your head is slightly smaller than the force of gravity on your feet, since your feet are closer to Earth's core. The distance to Earth's core is 3960 miles. Thus, the relative difference of gravity between your feet and your head is negligible. But this is not necessarily so for a black hole.

Suppose an astronaut decided to visit a stellar mass black hole, one with a mass of 10 solar masses. We will assume that the angular momentum of this object is relatively low, so that we may use the Schwarzschild metric as an approximation. To save fuel, the astronaut drops into the marginally-bound orbit at $2R_S$. The ship is in orbit, which is a free-fall trajectory, so the passenger feels no *net* force of gravity. But he will experience tidal forces. Suppose the astronaut is 6 feet tall, 70 kg in mass, and standing upright. Tidal forces will tend to stretch his head away from his feet, and compress him horizontally. But how strong are these forces? What would be the tidal force between his head and feet?

At 10 solar masses, the Schwarzschild radius (16.1) is:

$$R_S = \frac{2GM}{c^2} = 29.54 \text{ km} = 18.4 \text{ miles}$$

The ship orbits at twice that distance; r = 59.08 km. The tidal force (ΔF) will be the difference between the pull of gravity from the astronaut's head and the pull of gravity from the astronaut's feet. The distance between the astronaut's head and feet is Δr = 1.829 meters. From Newton's law of gravity, we derive the tidal force formula:

$$\Delta F = \frac{2GMm}{r^3} \Delta r \qquad (16.2)$$

$$\Delta F = 1.65 \times 10^9 N$$

Converting from the metric system, this works out to about 370 million pounds! That is, the feet of the astronaut would be pulled away from the head by this force. This of course would not be a pleasant experience for the astronaut. No one could survive the tidal forces so close to a stellar mass black hole. You would be stretched apart long before you even got close.   Astrophysicists refer to this stretching process as "spaghettification."

If you want to come close to the event horizon of a black hole and not be spaghettified, you must pick a supermassive black hole. The black hole at the center of our galaxy is estimated to be 4.5 million solar masses. This puts the Schwarzschild radius at 133 million kilometers. That's nearly the radius of Earth's orbit around the sun! At the marginally-bound orbital distance of $2R_S$, the tidal force between head and feet for our astronaut would be 0.065 newtons, or 0.014 pounds. The force would be hardly noticeable.

It may seem counterintuitive at first that the tidal forces would be weaker for the more massive black hole. But this is because tidal forces are due to the relative difference in distance to the singularity. The *percentage* difference in distance to the singularity between head and feet is much less for a supermassive black hole, because we must orbit at such a great distance.

Orbiting the supermassive black hole at $2R_S$, this astronaut will be time dilated relative to the distant stars. Some of this will be due to gravitational time dilation; the ship is near the source of gravity. The rest will be due to velocity time dilation; the ship orbits at a substantial fraction of the speed of light, as seen by a distant observer. At this special orbit, both effects happen to be exactly equal; each contributes $\sqrt{2}$ or a factor of about 1.414, such that the total time dilation for this orbit is exactly 2. So he could watch the universe move at twice the normal rate, and the distant stars would be strongly blueshifted.

If the astronaut thrusts in the opposite direction that he orbits, his orbit will decay into a lower one. As he descends to $1.5R_S$, he will enter the photon sphere. Any light emitted from his own ship will circle around the black hole and back to his ship, so that he will see multiple images of himself. In principle, photons can get stuck at this distance since they can orbit in a perfect circle here. But recall that orbits inside $3R_S$ are unstable. So in reality, even the slightest deviation from a perfect circle will eventually result in the photon leaving that orbit, either escaping the black hole entirely or else falling in.

Once reaching the photon sphere, the astronaut will have to fire his engines continually toward the black hole so that the reaction force will maintain his ship at this distance. Unassisted orbits are not possible inside the photon sphere. The universe would appear extremely distorted at this distance. The astronaut sees nothing below him. Above and to the front, the astronaut sees a bright and highly blueshifted star field. The star field seems to "split" into multiple mirror images at certain places, as if he were on the surface of a distorting mirror. It would look very strange.

Tidal forces are still negligible as the astronaut descends into a tighter, thrust-assisted orbit. But he certainly will feel the thrust of his own ship, as it accelerates against the inertial space falling in all around him. The universe above him continues to shrink, while the hole below creeps up like a black curtain. Time dilation increases, and the astronaut watches a highly blueshifted universe act as if in fast-forward.

Perhaps it is now time to think about switching the engines to full thrust, before the ship gets too close to the event horizon. But it is already too late. The ship passed through the event horizon some time ago. Nothing special happened, and the astronaut really didn't notice. From inside the event horizon, he can continue to see incoming light from the universe beyond. But he will never be able to share this view with anyone else. Any signal the astronaut attempts to send from inside the event horizon will always remain inside the event horizon. Inertial space falls toward the singularity faster than the speed of light, and thus no outgoing signal can escape.

At some point deep inside the supermassive black hole, tidal forces become noticeable. A previously-subtle upward pull on the head, and downward tug on the feet, becomes quite strong. The astronaut might try to reduce this effect by curling himself into a compact ball. This reduces Δr by roughly half. But it is a temporary reprieve. The inward horizontal compression and outward vertical stretching will eventually overcome his strength, and the astronaut will sadly become spaghettified, along with his ship. This will happen before the mass strikes the singularity, adding to its mass, and causing the event horizon to grow ever so slightly.

White Holes and Wormholes

In addition to black holes, which are real, general relativity allows for two theoretical objects that likely do *not* actually exist in the real universe: white holes and wormholes. Recall that the event horizon of a black hole is discovered by setting the escape velocity equal to the speed of light. The escape velocity is the velocity that an initially-stationary object would have if dropped from an infinite distance. We can think of inertial space as freely falling toward a mass at the escape velocity.

But escape velocity can also be outward-directed. Imagine watching a video of the infalling object *backward*. You would see an outgoing object. An object launched from Earth with exactly the same gravitational potential energy as kinetic energy will travel at the escape velocity at every distance in its journey. It will ever slow, but never stop, only becoming stationary in the limit that it reaches an infinite distance from the Earth. Although the object's direction of its velocity is opposite the direction of its acceleration, it is nonetheless in freefall. Therefore, it is locally inertial.

We think of a black hole as corresponding to infalling inertial space. But the very same metric also describes *outward* falling inertial space – as if space is being ejected from the mass at the escape velocity, slowing with increasing distance but never quite stopping. This makes sense because the formula for escape velocity involves a square root; and we may take either the positive or negative root. Both describe an object being pulled by gravity. But one object is inward-moving at escape velocity, whereas the other is outward-moving at escape velocity. Each resembles the backward-in-time movie of the other.

Thus, a white hole is a time-reversed black hole. Or, a black hole is a time-reversed white hole, if you prefer. Both are mass that is compacted to a singularity. Both are surrounded by an event horizon – a point of no return. But whereas an object may never escape the event horizon of a black hole, an object may never *enter* the event horizon of a white hole. And whereas mass may freely enter a black hole, it may be freely expelled from a white hole. Thus, white holes might emit radiation and appear to glow, just as black holes absorb radiation, and are black.

Make no mistake; both black holes and white holes have *positive* (attractive) gravity. This might seem strange at first, since white holes are always expelling mass and radiation. But such objects are expelled only because they have tremendous outward velocity – the "force" of gravity is still inward and constantly slows their outward movement. Time-reversed gravity is still attractive.

Another way to realize this is to consider an orbit of some object far away from a black hole. The object orbits because it is *attracted* (not repelled) by the gravity of the black hole, which makes its path deviate from a straight line. However, when the movie is played backward, the object merely orbits the (white) hole in the opposite direction. The orbit implies that gravity is still attractive.

It is counterintuitive (to me at least) that you can never enter a white hole, even though the force of gravity near one is strongly attractive. But this is because inertial space is being expelled *away* from the hole at the speed of light at the event horizon. Inside the event horizon, inertial space moves away from the singularity faster than the speed of light;

therefore, nothing between the event horizon and singularity can remain stationary. Anything inside the white hole must be expelled outward, just as anything inside a black hole must fall further inward.

We have good reason to believe that stellar mass black holes form when massive stars collapse in on themselves. But once a black hole forms, it can never lose mass (quantum effects notwithstanding). It can only grow larger as additional mass falls in. But it can never lose mass since nothing can ever escape the event horizon once inside. Thus, black holes can form, but cannot "unform."

Given that a white hole is a time-reversed black hole, it follows then that white holes can "unform," but can never form. And if they can never form, then it should hardly surprise us that none exist. Unless God created the universe with white holes already in it, and unless they have not yet expelled all their mass, then there cannot be white holes in the universe since they cannot form.

This seems to be a distinguishing feature of *time's arrow*: the observation that the future is perceptually different from the past even though most of physics is time reversible. After all, orbits work perfectly well backward. You cannot tell by watching any orbital paths or trajectories whether the movie is playing forward or backward. So why is it that the human brain (which apparently uses the laws of physics) can remember the past, but not the future?

Black holes seem to be one of those interesting phenomena that distinguish the direction in which time flows. Singularities may exist in the future state for matter, but not in the past state of matter. Although the mathematics of general relativity permits the existence of white holes, the real universe apparently does not. It would seem that only inward-directed free-fall inertial frames surrounding a mass are permitted to actually exist. This may be related to our perception of time and the fact that we are able to distinguish future from past events. In any case, there is no evidence that white holes exist anywhere in the cosmos.

A related concept is that of a *wormhole*. Consider the (2-dimensional) surface of an apple as representing the (3-dimensional) universe. If the radius of the apple is r, then an ant would have to walk a distance of roughly πr, to get from one side of the apple to the other. But a worm might eat straight through the middle. The worm gets from one side to the other by taking a shortcut straight through the apple, a distance of $2r$ which is less than πr.

Likewise, a wormhole is a hypothetical shortcut through space. Rather than traveling through normal space (like the surface of an apple), a wormhole could connect two distant points in space by an internal shortcut. By stepping through the wormhole, you might end up almost instantly in a distant galaxy. This doesn't violate the condition that mass cannot reach or exceed the speed of light since the mass did not move through space faster than c. Instead, the mass moved at less than c through a "short-cut" in space.

The equations of relativity allow the existence of a wormhole by connecting a black hole event horizon to a white hole event horizon in another universe, or in another location in the same universe. This is called an Einstein-Rosen bridge. Such a wormhole would not be traversable, much to the disappointment of sci-fi fans. Any object entering the black hole must eventually intersect the singularity. And no object can enter the event horizon of a white hole. So no journey from one end of the wormhole to the other would be possible.

Moreover, we saw previously that white holes do not seem to exist. Furthermore, the Einstein-Rosen bridge presupposes the *maximally extended Schwarzschild solution*, which is a temporally eternal version. In other words, it requires the black hole / white hole pair to have existed forever in the past and forever in the future. It is a fun mathematical construct. But since time has a beginning, this solution is disallowed in the real universe.

Other types of wormholes have been proposed, and some of them are theoretically traversable. However, in all cases they are apparently unstable, collapsing before any particle could make use of the short cut. Some physicists have invoked "exotic matter" to stabilize the wormhole. Exotic matter is a hypothetical substance with negative rest mass / energy. However, physicists have not discovered any such material. And there are additional problems with wormholes, such as the introduction of temporal paradoxes. This is because a wormhole might allow for the possibility of time travel into the past.

A wormhole connects two points in space, so that when a person enters one end of the wormhole, he exits at the other end. But what if we moved one end of the wormhole rapidly through space – say out to a distant star and then back? Due to time dilation, it would not age as much in external space as the other end, just like the twin paradox. But the two ends are still connected internally. So an observer who enters the accelerated wormhole end would exit at a much earlier time than he left. Wormholes would therefore make it possible to travel back in time, which can lead to temporal paradoxes.[23] This again suggests that wormholes do not exist in the real universe.

Physicists and mathematicians may enjoy considering these interesting solutions to Einstein's field equations. But the real universe must obey not only the rules of general relativity, but other physical laws as well, such as thermodynamics and quantum mechanics. While wormholes and white holes may be permitted by relativity, they appear to be disallowed by other branches of physics. Thus, we do not expect to find them in the real universe. Black holes, however, are real.

---

[23] Some physicists have suggested that quantum effects would destroy any such time travel wormhole. The idea is that virtual particles would accumulate in the wormhole with progressively increasing intensity, creating a feedback loop that would destroy the wormhole before any information-bearing material could pass through it.

Chapter 17

The Curious Case of the One-Way Speed of Light

Returning to the realm of special relativity, we now consider a very interesting phenomenon involving the concept of *one-way speeds*. In particular, we want to know the one-way speed of light. This requires some explanation, but the results are truly baffling. Most attempts to measure the speed of light actually measure its *two-way* or *round-trip* speed. Consider the following thought experiment.

At location A, we have a flashlight and a very accurate atomic clock. At exactly noon, the clock activates the flashlight, sending a pulse of light down a long hallway to point B. For the sake of making the math easy, let's suppose that B is 186,282.3974 miles from point A. At point B is a mirror, which reflects the light beam, sending the pulse back to A. A sensor at point A records the atomic clock time when it receives the reflected light. Given this setup, the clock would record exactly two seconds between when the light was sent and when it returned.

The speed of light is calculated by taking twice the distance of A to B (since the light had to travel this distance twice – from A to B and back to A) and dividing by the time difference (between start and finish) recorded by the atomic clock. When conducted in a vacuum, such experiments always give the speed of light as 186,282.3974 miles per second. The light had to travel twice the distance of 186,282.3974 miles, and it did so in *two* seconds.

But this is a round-trip, time-averaged speed – meaning the light had to travel from A to B *and then back to A*. There is no guarantee that the light traveled at this speed *throughout the journey*, though many people might assume such. In other words, people tend to assume that the light took one second to go from A to B, and one second to return to A. But this is an *assumption* – not something that was measured in the experiment.

Suppose for the sake of argument that the light took 1.5 seconds to go from A to B, and only 0.5 seconds to return to A. Clearly, the light would be slower on the outbound journey, and faster on the return trip. Yet its *average* speed would remain unchanged. The clock would only record two seconds for the *entire* trip. Dividing the total distance by the total time, we again find the *average* speed of light to be 186,282.3974 miles per second. Yet the light was slower than this on the outbound trip, and faster on the return trip.

Alternatively, the light might take no time at all to zip from A to B, and take 2 seconds to return to A. Again, our clock would measure two seconds. Dividing the total distance by the total time, we find that the average speed of light is again 186,282.3974 miles per second. The reverse scenario gives the same answer: if the light takes 2 seconds to go from A to B, but takes no time at all to return to A, the average speed still works out to 186,282.3974 miles per second.

"But why would the speed be different from A to B than it is from B to A?" I don't know of any reason why they should be different. But then again, I don't know of any reason why they would have to be the same. People might emotionally prefer the symmetry of having the speed from A to B be the same as the speed from B to A. But does preferring something automatically make it so? Should we expect the universe to conform to our emotional preferences?

If there is one primary truth that we learn from the physics of Einstein, it is that the universe does not always conform to our preferences or expectations. Few people would prefer or expect that motion affects the passage of time. But it does nonetheless. Hardly anyone would anticipate that mass increases with velocity; but it does. The universe is under no obligation to function according to our preferences or expectations. Sometimes it does. But sometimes it does not.

Therefore, we may not simply assume that the one-way speed of light is the same from A to B as it is from B to A. That may turn out to be the case; but it is not rational for us to make such an arbitrary assumption without evidence. Nor is it rational to assume something simply on the basis that no one has proved its opposite. In logic, that would be called the fallacy of the appeal to ignorance.[24]

Therefore, if we are going to be rational and scientific, we may not arbitrarily assume that the speed of light is the same going from A to B as it is going from B to A. This might turn out to be the case, but we currently have no evidence for or against the claim. To be scientific, we must perform an experiment to test the claim that the speed of light is the same from A to B as from B to A. That is, we must measure the speed of light on a one-way trip.

Attempts to Measure the One-Way Speed of Light

So how might we measure the one-way speed of light? We wish to know if the light took the same time to go from A to B as from B to A. Clearly, we will need *two* clocks for this experiment. One clock resides at point A. It triggers the light pulse at exactly 12:00. The other clock resides at point B. It records the time when the light arrives. We simply subtract the start time (12:00) from the time recorded at B when the light arrives to get the time of this one-way trip. We know the distance is 186,282.3974 miles. Dividing distance by time gives us our answer. Simple, right?

Unfortunately, there is a horrendous problem with the above scenario, as illustrated by a similar scenario. Suppose that Jeff tries a scaled-down version of this experiment in his office. He doesn't have the long hallway of course, but he does have the distance between his wrist watch and the clock on his desk: a distance of about five feet. When his watch strikes noon, Jeff turns on a flashlight pointed directly at the clock on his desk. His friend Dan is watching the desk clock, and is ready to record the time when the light strikes this clock. He reports that the clock reads 12:05 the instant the light arrived.

The difference in time would be five minutes, and the distance is 5 feet. Dividing the distance by the time gives a meager velocity of one foot per minute. People walk about a hundred times faster than this. So clearly this can't be right. Jeff and Dan would have to be rather dim to think that they had really measured the speed of light at one foot per minute. Yet, the experiment could actually occur with those numbers. What is going on?

---

[24] The appeal to ignorance is the error in assuming that something must be true by virtue of the fact that no one has (yet) proved it false. This is fallacious reasoning because something might be false even if people have not yet discovered the evidence that disproves it.

Clearly, the clock on the desk is ahead of Jeff's wrist watch by about 5 minutes. There is no strange time dilation effect here. Rather, either Jeff's watch or the desk clock or both are not set to the same time. They have not been *synchronized*. They do not read the same time at the same time. This would be obvious if Jeff would place his wrist watch next to the desk clock. They would read different times at the same time. Thus, we learn an important principle:

To measure the one-way speed of light requires clocks separated by a distance to be exactly synchronized.

Hopefully, this is quite obvious. To measure the velocity requires us to divide the distance traveled by the time difference. But in order for the time difference to be accurate, the clocks must be synchronized – they must *read* the same time *at* the same time so that when one clock reads noon, so does the other clock at that very instant.

Not only do the clocks need to be synchronized, but they must be *exactly* synchronized. If clock B is only one second behind clock A, then this could make the difference between measuring a one-way speed of 186,282.3974 miles per second, and *infinity*! Since light is very fast, our clocks need to be *exactly synchronized*.

Synchronizing Distant Clocks

Some people might respond, "Okay. So we'll synchronize our clocks before the experiment begins. No problem." But this turns out to be much harder than what we might assume. Synchronizing two clocks that are at the same location is easy. Just set the time on one to match the other. We can see that the clocks are synchronized because they read the same time at the same time. But when two clocks are separated by a large distance, how would you synchronize them?

Normally, clocks on Earth are synchronized by radio transmission. A radio transmitter at location A sends out a radio pulse at noon. The clock at B receives the radio pulse and sets itself to noon. Many people own clocks that do just that – they receive a radio transmission from the atomic clock in Boulder, Colorado, and automatically adjust their time to match. And now we have synchronized clocks, right?

Not exactly. Radio waves are very fast, but they are not necessarily instantaneous. It may have taken a small amount of time for the radio pulse at A to travel to B. By the time B receives the pulse and sets itself to noon, the clock at A is already slightly *past* noon. Normally, we don't worry about this slight difference. But if we are going to measure something as fast as light, "almost" synchronized is not good enough. Even the slightest discrepancy will introduce enormous error into our velocity estimate. We need clocks that are *exactly* synchronized. Radio transmissions don't quite cut it.

Of course, if we knew how long it took the radio pulse to travel from A to B, we could compensate for this. Suppose, for example, that it takes one tenth of a second for the radio pulse to travel from A to B. The pulse is emitted when clock A strikes noon. When clock B receives the signal, instead of setting it to noon, we set clock B to 0.1 seconds *past* noon. That will work – if we knew how fast radio travels.

So how fast does radio travel? Interestingly, radio travels at the same speed as light. But we don't know the speed of light on a one-way journey; that's the question we're supposed to be answering! If we don't know the one-way speed of light, we don't know the one-way speed of radio either.[25] And that means we don't know how long it took the radio pulse to get from A to B, and therefore we don't know how far ahead of noon to set clock B when it receives the signal.

It is a catch-22. We would have to know in advance the one-way speed of light (and thus the one-way speed of radio) in order to synchronize clocks by radio transmission, so that we could then measure the one-way speed of light. But if we already knew the one-way speed of light, there would be no need to perform this experiment. And since we don't know the one-way speed of light, we have no way of synchronizing two clocks by radio transmission.

Some people have proposed a variation on the radio-synchronization method. They suggest putting the radio transmitter exactly halfway in between A and B at point M. They reason that any radio pulse emitted by M will arrive at point A at the same time it arrives at point B since the distance is the same. Granted, clocks at A and B might both be behind a clock at M, but at least they will be synchronized to each other. Right?

Well, no. This method assumes that radio traveling from M to A propagates at the same one-way speed as radio traveling from M to B *in the opposite direction*. But the very claim under investigation is whether

---

[25] We do know that the one-way speed of radio is the same as the one-way speed of light (even if we don't know what that speed is). If a radio pulse and light signal are sent from A to B, both will arrive at B at the same time. This type of experiment has been performed.

light (and thus radio) in different directions travels at the same or different (one-way) speeds. We cannot arbitrarily assume that it does in our experiment to find out if it does. That would beg the question.

As another method to synchronize clocks, some people have suggested that we synchronize the two clocks at the same location M and then move one to A and move one to B. It is easy to see if two clocks are synchronized when they are both at M; we can just look at them because there is no light travel time. So we move them to A and B, and they are still synchronized – right?

Well, no. That is not necessarily so. We learned in an earlier chapter about time dilation. We understand that motion affects the passage of time. So the very act of moving the clocks away from M has affected the rate at which they tick. Each clock has experienced some time dilation, and so there is no guarantee that they are *still* synchronized. After all, moving clocks tick slow.

Perhaps all is not lost. We derived the formula for time dilation in an earlier chapter. And so if we know the speed at which the clocks were moved to A and to B, we can calculate how much time dilation each clock experienced during its journey. We can then adjust each clock to compensate for the effect of time dilation, so that they are once again synchronized. In fact, if the clocks are moved at the same speed, our time dilation formula predicts that they will experience the same time dilation. So any adjustment will be unnecessary. They will still be synchronized to each other. Correct?

Again, we have overlooked something. In the formula for time dilation (when we expand the Lorentz factor) is the constant "c." And what does that letter represent? Ah, yes. It is the speed of light. The formula for time dilation requires us to know in advance the speed of light – the very thing we are attempting to measure. If the one-way speed of light is different from M to A than it is for M to B, then it stands to reason that *time dilation* – which depends on the speed of light – will also be different for clock A and clock B. This will be the case even if the clocks moved at the same velocity v, but in opposite directions.

In fact, in our derivation of the time dilation formula, we tacitly assumed that the one-way speed of light is the same in all directions and is also the same as the round-trip speed – namely 186282.3974 miles per second. But now we are questioning whether that is in fact the case. If it isn't, then the *actual* time dilation formula may be quite different from the one we derived earlier. It will likely have directional dependence. So we may not assume that clocks at A and B are still synchronized even if they have traveled at the same speed from M, since they have traveled in different directions.

The point is this: we don't (at present) know how much time dilation occurs if the one-way speed of light has a different value in different directions. And yet, we would need to know how much time dilation clocks at A and B have experienced, so we can then adjust them (if necessary) to be synchronized, so that we can then measure the one-way speed of light. Once again, we are in a catch-22. We would need to know in advance the one-way speed of light, in order to derive the correct time dilation formula, in order to synchronize the clocks at A and B, in order to measure the one-way speed of light.

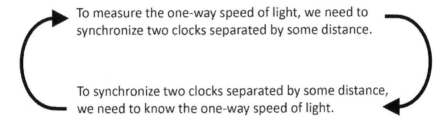

To measure the one-way speed of light, we need to synchronize two clocks separated by some distance.

To synchronize two clocks separated by some distance, we need to know the one-way speed of light.

Slow Clock Transport

There is a modification of the above method that we might try. We know that time dilation is very small for ordinary velocities because we can see that moving clocks tick at about the same speed as stationary ones in our everyday experiences. So what if we synchronized the two clocks at M, and then moved them to A and B *as slowly as possible*? Perhaps by moving the clocks at very slow speeds, time dilation will become negligible, and the clocks at A and B will remain synchronized.

This method is called "slow clock transport." It is very tempting to think that this is the answer. But when we examine it closely, there is a flaw. Suppose we synchronize clocks at M, and then quickly move one of them to A *and back to M*. We know from the postulates of relativity that the moving clock will be slightly behind the clock at M. It will have experienced some time dilation. (This is provable regardless of the one-way speed of light, though we will not show the proof here.) But, if we instead moved it *slowly* to A and back, it will be *almost* synchronized with the clock at M — only slightly behind.

The slower the velocity of the moving clock, the less it will be behind the clock at M when brought back together. In fact, it is possible to prove that in the limit as velocity goes to zero, the two clocks will be exactly synchronized when brought back together. In other words, if you synchronize two clocks at M, and then move one of them to A and back to M *as slowly as possible*, it will still be synchronized with the stationary clock at M.

So perhaps we have our answer. We simply synchronize two clocks at M, and then separate them *as slowly as possible* to positions A and B. They should still be synchronized. Right? After all, we know that if we slowly brought them back to M, they would indeed be synchronized.

As always, there is a catch. Namely, how do we know that the clocks *remained* synchronized during the entire journey? We only know that they were synchronized initially, and are still synchronized when brought back together. But this does not guarantee that they *remained* synchronized the entire time.

In other words, perhaps the clock gained a second when it was moved to A and then lost a second when moved back to M. This would be consistent with our observations. Conversely, perhaps the other clock lost a second when it was moved to B, but then gained a second when it was moved back to M. In this case, both clocks would read the same time when they are brought back together, *yet their time was different by two seconds when they were separated.* In fact, it is possible to prove from first principles that this is exactly what happens if the one-way speed of light varies in different directions. We will show the equations below.

Slow clock transport does not guarantee that the clocks have *remained* synchronized throughout their trip. It only guarantees that they become synchronized again when brought back together at very slow speeds. Thus, moving clocks as slowly as possible from M to A and B does not ensure that the clocks at A and B are still synchronized. Slow clock transport tacitly assumes that the one-way speed of light is the same in all directions.

It is again a catch-22. In order to measure the one-way speed of light, we need exactly synchronized clocks at A and B. Yet the very act of moving clocks to A and B causes them to desynchronize. And we cannot compensate for this, because the amount of time dilation *depends* on the one-way speed of light – the very thing we do not know. So neither signal transmission, nor moving clocks, can guarantee that the two clocks at A and B are synchronized, so that we could then measure the one-way speed of light.

Other Attempts to Synchronize Distance Clocks

People have proposed some very elaborate thought experiments in order to synchronize distant clocks to measure the one-way speed of light. But they are all merely variations on one of the two methods above. Hence, they all fail for exactly the same reason. That is, the clocks are supposedly synchronized by the transmission of a signal; but the signal speed invariably depends on the (unknown) one-way speed of light. Or, the clocks are moved in some fashion and assumed to be synchronized, but without knowing the correct time dilation formula, which depends on the one-way speed of light.

For example, some people have suggested using an electrical signal to synchronize clock B to clock A. But electricity is not quite instantaneous. In fact, electricity travels at the (one-way) speed of light. So clock B will be slightly behind A by an unknown amount. Electricity is fine if we want the clocks to be *approximately* synchronized. But it is not good enough to measure the one-way speed of light.

Perhaps we could use sound waves to synchronize the two clocks. Sound is much slower than light, and depends on air temperature (substantially) and pressure (slightly), and so this would be hard to implement. But theoretically, we could measure the speed of sound, and use a sound signal to synchronize clock B with clock A, making adjustments at B for the time lag. But is the speed of sound the same in all directions?

Sound is caused by compression of molecules as they collide with each other. But the collision of molecules is due to their electromagnetic interaction. And electromagnetic action propagates at (you guessed it) the one-way speed of light. Therefore, if the one-way speed of light is not the same in all directions, then the speed of sound will also be slightly different in different directions – enough to throw off the synchronization of clocks to the point that it invalidates the experiment.

In similar fashion, all other signaling mechanisms depend directly or indirectly on the one-way speed of light. Using a signal other than light might make it more difficult to spot the catch-22. *But it is always there.* It is all too easy to make an assumption that tacitly *assumes* that the one-way speed of light is the same in all directions.

Even very sharp minds are sometimes snared by circular assumptions. I recall an experiment proposed in peer-reviewed technical literature that purported to measure the one-way speed of light.[26] As a

---

[26] http://arxiv.org/abs/1310.1171

simplified explanation, essentially, it used a long shaft with two circular disks, one at each end of the rod. The center of each disk goes through the rod, so that when we rotate the rod, each disk turns at the same rate. Imagine a small hole punched into each disk at the same location. A collimated light beam is set up so that it shines through the hole in disk A, and then onto the hole in disk B, and finally is received by a detector at C. Obviously, A B and C are in a straight line.

The rod is connected to a motor, and begins rotating at high velocity. So the disks rotate with the rod. Most of the time, the light beam is blocked by disk A. But once per revolution, the holes in the disks line up with the light beam, and the detector receives the signal. But if the one-way speed of light is not quite instantaneous, then it will take a little bit of time for the light beam to travel through the hole at A to reach disk B. And if the disks are spinning fast enough, by the time the light beam reaches B, the hole will no longer be lined up with C. And so, at a certain rotational speed, no light will ever reach the detector. By knowing the size of the holes, the distance between them, and measuring the rotational speed at which the detector no longer receives any signal, we can apparently calculate the amount of time it takes for the beam to travel from A to B – the one-way speed of light.

It's a very clever setup. But it fails. The reason is because the experiment very subtly assumes that the one-way speed of light is the same in all directions. And how does it assume this? The method presupposes that the rod is infinitely rigid – that it does not "twist" when rotated. But the rod is made of atoms which are held in place by chemical electromagnetic bonds. When we change the position of these atoms by rotating the rod, the electromagnetic bonds travel at the one-way speed of light. Now if the one-way speed of light is the same in both directions, then no twisting will occur due to symmetry. That is what the experimenters assumed. But if the speed of light is faster in one direction, then *twisting will occur*.

Suppose for example that the light from A to B is instantaneous, and is slower from B to A, then the rod will twist such that disk A is rotated slightly ahead of disk B at the same instant. So at a certain rotational speed, the holes will no longer align due to the twisting of the rod, and the detector will no longer receive a signal at that rotational speed – even if the light is instantaneous. When we go through the math and calculate what rotational speed is necessary for the twisting of the rods to block the light beam, we find that it is exactly the same speed that the non-twisted rod reaches when the one-way speed of light is assumed to be c.

In other words, no matter what the one-way speed of light is, this experiment will always give exactly the same result. It therefore cannot distinguish between different values of the one-way speed of light. In neglecting to account for any twisting of the rod, the experimenters effectively assumed the that the one-way speed of light is symmetric – the very thing they were supposed to be proving.

I mention this illustration not to disparage the people who implemented it, but rather to point out that it is very easy to make an assumption that tacitly assumes that the one-way speed of light is c. Interestingly, this experiment is capable of measuring the round-trip speed of light, even though the actual light beam only makes a one-way trip. We will discover the reason for this strange result later in this chapter.

Historical Measurements of the Speed of Light

Interestingly, the first successful measurement of the speed of light *seems* to measure the one-way speed. But when we break down the details, we find that it actually measures the two-way (round-trip) speed of light. This experiment was performed by the Danish astronomer Ole Rømer in the 1670s. He, along with other astronomers, precisely measured the timing of eclipses of Jupiter's moons, particularly Io. He noticed that the eclipses took place a bit earlier when Earth was closer to Jupiter than they did when Earth was farther from Jupiter. He reasoned that this was merely an optical effect due to the fact that the light had to traverse a greater distance when Earth was farther from Jupiter.

In other words, it takes light (from Jupiter) a little over 8 minutes to cross half the diameter of Earth's orbit – about 93 million miles. The modern value would be 8.317 minutes. Given that the light takes 8.317 minutes to travel this distance (92.9558 million miles) gives a speed of about 186,000 miles per second. But this eight minutes is the time it takes for light to travel one-way, from a point where Earth is closer to Jupiter to a point where Earth is farther from Jupiter.   Did Rømer actually measure the one-way speed of light?

It certainly seems so. Yet, there is always a hidden assumption. In this case, Rømer assumed that clocks on Earth tick at a constant rate, regardless of whether Earth is moving toward or away from Jupiter. In other words, he neglected to account for the effects of time dilation. Rømer had merely *assumed* that the apparent advancement or retardation of the eclipses was due to the extra distance that light had to travel. It didn't occur to him that Earth's motion through space could cause clocks to tick at different speeds. He had wrongly assumed that time is absolute, and not affected by velocity.

He could certainly be forgiven for making the assumption of absolute time since the physics of Einstein had not yet been discovered. Everyone at that time thought that time was absolute and unaffected by motion.   But today we know better.   As the Earth orbits the sun, it is sometimes moving toward Jupiter, and sometimes moving away from Jupiter.   And this motion affects the rate at which clocks on Earth tick relative to some hypothetical distant observer. Furthermore, the degree of time dilation depends on the one-way speed of light. So we cannot know whether the time lag in the eclipses of Jupiter's moons is due to (1) the increasing distance that light must travel, (2) the effect of time dilation, or (3) some combination of the two.

But again, we note the remarkable result that Rømer's attempt to measure the one-way speed of light actually does give the correct value for the *round-trip* speed of light. It is truly astonishing that a conceptually flawed experiment gives the right answer to the wrong question. There is a mathematical reason why this is so, and it is something we will cover in a later section.

The Significance of the One-Way Speed of Light

Physicists have discussed and debated the issue of the one-way speed of light for the past century. Yet, no one has been able to come up with a genuine experiment that can in principle measure the one-way speed of light without tacitly assuming the one-way speed of light. Several experiments have been proposed in the technical literature. But inevitably they are flawed, and are exposed as such by another technical paper.

This seems to indicate something profound about the universe. Apparently, nature is completely unaffected by the one-way speed of light. Only the round-trip speed matters when it comes to observable, testable, repeatable results. But what does this mean?

First, it means that the previous chapters of this book have not been a waste of time. Prior to this chapter, we have tacitly assumed that the one-way speed of light is always c regardless of direction. This may be wrong, and yet it still always gives the right answer as to what we will *observe* in any testable, measurable experiment.

Second, it means that we *could* have picked another value for the one-way speed of light, allowing light to travel at different speeds in different directions. And all of the equations we derived on that basis would also correctly describe the observable outcome of any experiment. For example, we could allow light to be infinitely fast when traveling from west to east, and ½c when traveling east to west. This produces a time-average round-trip speed of c. Had we made this assumption, the resulting equations might look a bit different, but they would still give the correct prediction about what we will actually observe in any given experiment. To put it colloquially, *nature doesn't care what value we assume for the one-way speed of light.*

Let me clarify what this means. Suppose we are tracking a nearby asteroid as it travels through space, and we want to know if it will eventually collide with the Earth. Emily and Jake are both tracking this asteroid, keeping track of its position as a function of time. Jake uses the standard assumption about the one-way speed of light that we have used in this book; namely he supposes that the one-way speed of light is the same in all directions. Emily, however, supposes that light travels infinitely fast when going from west to east, and light travels at only ½c when traveling from east to west.

Each observer uses his or her presupposed one-way speed of light to synchronize distant clocks by radio transmission. Naturally, they will disagree on whether any two given clocks are synchronized. Consequently, they will record somewhat different times for any given location of the asteroid. That is, Emily claims that the asteroid reaches location M at time 12:33:47, whereas Jake claims the asteroid reaches location M at time 12:45:02. Furthermore, Jake's estimate for the asteroid's speed may differ slightly from Emily's, since each observer uses a different standard to mark time. But they will always agree on the objectively measurable outcome – whether the asteroid will in fact strike the Earth. Assuming that both do their computations correctly, they will agree on the outcome, even though they use different definitions of time, velocity, and the one-way speed of light.

By analogy, Earth observers in California might claim the asteroid strikes Earth at 2:45 pm, whereas observers in Ohio claim it strikes at 5:45 pm. Ohio is on Eastern Time, whereas California is on Pacific Time. Consequently, observers from each time zone will disagree on the time when an event occurs – but they all agree *that* it occurs. Likewise, special relativity always gives the same answer to objectively measurable phenomena regardless of any assumptions about the one-way speed of light. The one-way speed of light merely sets how we choose to synchronize clocks – how we define "now" much like time zones on Earth. This has been known for some time.

In 1970, John Winnie showed that the measurable effects of special relativity do not depend on any particular value of the one-way speed of light – only the round-trip speed. He found that whatever the one-way speed of light turns out to be, it has absolutely no effect on any objective observations or measurable outcome of anything in the universe. It only affects how we define "simultaneous" and thus how we time stamp various events. It is instructive to review Winnie's research.

The Epsilon Equations

John Winnie's two-part paper is entitled "Special Relativity Without One-Way Velocity Assumptions" (Part I and Part II). It was published in the journal *Philosophy of Science*, Volume 37, No. 1 and 2. In these documents, Winnie uses the symbol c to denote the round-trip speed of light: namely, 186,282.3974 mi/s. The one-way speed of light is allowed to differ from this value depending on the direction of travel. Following the usage of Reichenback, Winnie used the symbol epsilon (ε) to denote any asymmetry in the one-way speed of light.

When $\varepsilon = 0$, this represents outward-directed light moving at infinite speed whereas inward-directed light moves at speed $\frac{1}{2}c$. The case where $\varepsilon = 1$ represents the reverse, where outward-directed light moves at $\frac{1}{2}c$ and inward directed light is infinitely fast. When $\varepsilon = \frac{1}{2}$, this represents the one-way speed of light being exactly c in both outgoing and incoming directions. So $\varepsilon = \frac{1}{2}$ is the "standard" value and the one we used in the previous chapters. But Winnie showed that relativity works just as well (always gives the correct answer) for any value of epsilon.[27] So he derived the basic equations of relativity, without making any assumption about the value of ε. We will not repeat his proofs here. But there is value in examining his results:

Without making any assumptions about the one-way speed of light (and thus the value of ε) Winnie derived the following formula for time dilation[28]:

$$T' = T \sqrt{1 - \frac{v^2}{c^2} \cdot \frac{c}{c + v(2\varepsilon - 1)}} \tag{17.1}$$

The first expression is familiar to is; it is T divided by the Lorentz factor. Notice that if $\varepsilon = \frac{1}{2}$, the expression after the radical reduces to one. So when $\varepsilon = \frac{1}{2}$ we find that this reduces to the standard time dilation formula:

$$T' = T \sqrt{1 - \frac{v^2}{c^2}}$$

Or equivalently,

$$T = T'\gamma$$

Winnie also derived the formula for length contraction, again leaving ε as a free parameter[29]:

---

[27] The value of ε is allowed to be anything between and including 0 and 1. Values below zero or above one still work mathematically. However, they would result in time stamping some effects earlier than their preceding cause, which appears to violate causality. Hence, we restrict the value of epsilon to be $0 \le \varepsilon \le 1$.

[28] Winnie had the moving frame traveling in the negative x-direction, whereas we have it moving in the positive x-direction. Thus, there is a sign difference in the velocity term between his convention and ours.

[29] Again, we select the opposite sign for v, since we define the primed frame to be moving in the positive x-direction, whereas Winnie had it in the negative x-direction.

$$L = L' \sqrt{1 - \frac{v^2}{c^2}} \cdot \frac{c}{c + v(2\varepsilon - 1)} \qquad (17.2)$$

As before, the formula is remarkably similar to the one we derived in an earlier chapter, but with an "extra" expression trailing the radical. Once again, when $\varepsilon = \frac{1}{2}$, the right-hand expression reduces to unity, and drops from the expression, leaving the familiar form of length contraction (5.1):

$$L = L' \sqrt{1 - \frac{v^2}{c^2}}$$

Or equivalently,

$$L = L'/\gamma$$

But when epsilon is not equal to $\frac{1}{2}$, the "extra" term on the right does not reduce to one. Thus, both time dilation and length contraction at a given velocity depend on the one-way speed of light. We will focus on this new epsilon-dependent version of the time dilation formula because the new "extra" term leads to some interesting consequences.

What consequences arise from using the above equations which make no assumptions about the one-way speed of light? First, the extra term gives the time dilation effect a quasi-*linear* (rather than quadratic) dependence on velocity. Notice that the velocity term in the denominator on the right-hand expression is v, it is *not* v^2. What this means in practice is that at low velocities the effects of time dilation are much stronger when $\varepsilon \neq \frac{1}{2}$ than they are when $\varepsilon = \frac{1}{2}$. Previously we found that velocity had to be as high as 14% the speed of light in order to have a meager 1% effect on time dilation. But this is for the case $\varepsilon = \frac{1}{2}$. If we set $\varepsilon = 1$, then velocity needs to reach only 1% the speed of light to have a 1% effect on time dilation. At 5% of c, the effect is just over 5% and so on.

Second, since the extra term is linear rather than quadratic, the effect is direction-dependent. When $\varepsilon = \frac{1}{2}$, the linear term disappears, and the situation is symmetric; so the time dilation of an object approaching at speed v is exactly the same as the time dilation of an object *receding* at speed v. But when $\varepsilon \neq \frac{1}{2}$, the situation is no longer symmetric; hence the time dilation experienced by an object approaching at speed v will *not* be the same as the time dilation experienced by an object receding at speed v.

Third, and again as a result of the linear term, it is possible for certain values of ε at certain velocities to have time *contraction*. That is, under certain circumstances, the moving clock can actually tick *faster* than the stationary one.   For example, when $\varepsilon = 1$, clocks approaching the stationary observer will tick faster than the observer's clock. (This effect is not possible for $\varepsilon = \frac{1}{2}$, under which moving clocks always tick slow regardless of direction.)

The Effects of the Choice of ε in Time-Stamping Events

In light of these details, it may seem at first that the value of ε will have a dramatic effect on what we observe and measure in the universe. But in fact, it doesn't.   Remarkably, it has no effect at all.   Two observers, regardless of their choice of ε, will always predict and observe the same result for any experiment. They may disagree on what time we should assign to an event because their choice of ε means that they have a different method of synchronizing clocks. But any choice of ε will predict the correct observable result when we use the full (epsilon-dependent) time dilation formula.

Let me clarify with an example. Assuming $\varepsilon = \frac{1}{2}$, Jake predicts that two asteroids will collide at 1:37 pm. He knows this because he has measured their positions and velocities, and is able to compute their projected path. He notices that both will be in the same place at 1:37 pm – hence a collision. Emily assumes that $\varepsilon = 1$. So she measures a different time and velocity for both asteroids because her assumptions about light travel time differ from Jake's. She computes their future paths, and notices that they will be in the same place at 1:45. Hence, she also predicts a collision.

Since Jake and Emily make different assumptions about the one-way speed of light, they disagree on the exact timing of some distant event. But they both agree on what happens – the asteroids will collide. *The choice of ε merely affects how they record the time of the event.* And there is no objective, reference-frame independent way to measure the time of a distant event because of the relativity of simultaneity, as discussed in chapter 7.

Consider the following scenario. Jake and Emily stand at location A. At precisely noon, they activate a flashlight, sending a beam of light toward a mirror at location B, a distance of 186,282.3972 miles away. The light then reflects back to A. Jake presupposes that $\varepsilon = \frac{1}{2}$, so he takes the one-way speed of light to be c in both directions. From his perspective, the light takes one second to get to B, and one second to get back to A. He will see the reflection at 12:00:02.

Emily presupposes that $\varepsilon = 1$, so she takes the outgoing speed of light to be $\frac{1}{2}c$, and the incoming speed to be infinite. From her perspective, the light takes two seconds to go from A to B, and no time at all to return back to A. She will therefore see the reflection at 12:00:02. Of course, that's exactly what Jake sees. It makes no difference which one-way speed we assume; the result that each observer sees is exactly the same.

Jake and Emily will disagree on what time the light reflected off the mirror at B because they disagree on the value of ε. Suppose that they wanted to figure out who is right. So they place a clock at B to register when the light arrives from A. But is this clock synchronized with the clock at A?

Emily decides to synchronize clock B with clock A by a radio transmission from A. Since she supposes that $\varepsilon = 1$, she believes that light (and therefore radio) takes two seconds to get from A to B. So she sends a radio pulse from clock A when it reads 11:59:58, just two seconds before noon. It will take exactly two seconds for the signal to arrive at B, at which time clock B will set itself to noon. According to Emily, the clocks are now synchronized.

But Jake objects to this procedure. In his view, $\varepsilon = \frac{1}{2}$, and therefore the light takes only *one* second to get from A to B, not two. He believes that Emily *should* have sent the radio pulse when clock A reads "11:59:**59**" – just one second before noon. In Jake's view, Emily sent the radio pulse one second earlier than she should have, and therefore clock B is not synchronized with clock A but is actually one second fast. It will be tempting at this point to ask, "Who is correct?" Is the clock at B synchronized with A as Emily claims, or is it actually one second ahead as Jake claims? Jake and Emily decide to do an experiment to find out.

The experiment is conceptually simple. The next day at exactly noon, they will bring clock A to the location of B and see if the clocks are synchronized or not. Of course, they know that motion affects the passage of time. But we now have the full equation for time dilation in terms of ε. So Jake and Emily will use the time dilation formula to compensate for the different rate at which clock A ticks while it is moving, and they will adjust it accordingly. Then, after that calibration, they will see if it really is synchronized with B. So when we go through the math, who is right?

>   At exactly noon according to clock A, Jake and Emily begin moving the clock at A to point B. At the moment the journey starts, Jake believes that clock B reads 12:00:01, since he believes it to be one second ahead of A. But Emily believes that clock B reads 12:00:00 – the same as A. Let's suppose that by Jake's measurements, the clock is moved at 10% of the (round-trip) speed of light. (So $v/c = 0.1$.) The journey will therefore take 10 seconds as measured by a stationary observer. Jake expects that when they arrive, clock B will read "12:00:11" since it is one second fast, and the journey will have taken ten seconds.
>
>   But clock A will have experienced some time dilation. From the formula, using Jake's assumption that $\varepsilon = \frac{1}{2}$, at 10% the speed of light, clock A ticks slow such that $T'/T = 0.994974$. This effect accumulates, so that during the ten-second trip clock A ticks 9.949874 seconds. By Jake's computation of the time dilation, clock A lost 0.0501256 seconds due to time dilation during the journey. So Jake will adjust clock A by setting it forward by 0.0501256 seconds, such that it now reads the correct time: "12:00:10." But clock B reads "12:00:11" at this moment. Jake concludes that he is right; clock B is indeed one second ahead of clock A. Or is it?
>
>   Since Jake believes light takes only one second to get from A to B, whereas Emily believes it takes *two* seconds, when Jake says that moving the clock from A to B takes ten seconds, Emily will measure this as *eleven* seconds.[30] But the time dilation the clock experiences will be greater in her view than in Jake's because she claims $\varepsilon = 1$. So from the full time dilation formula, we find that $T'/T = 0.904534$. Over the course of the *eleven*-second journey, clock A ticks only 9.949874 seconds. Emily concludes that the clock lost 1.0501256 seconds during the trip, and sets it forward by that amount so that it now reads "12:00:11" – exactly the same as clock B. She concludes that they were indeed synchronized.

---

[30] Consequently, she will take the velocity of the clock to be different from Jake's assessment. In her view, the clock travels at 9.0909% the speed of light. The choice of ε, therefore, affects all one-way speed measurements, not just light. Note, however, that the full-time dilation formula with ε-dependence uses the velocity v as seen by the $\varepsilon = \frac{1}{2}$ observer. In this case, that is Jake's assessment of the velocity. So $v = 0.1c$.

Notice that both Jake and Emily correctly predict what clock B will read at the end of its journey: 9.949874 seconds past noon. That is the only objectively measurable result. They disagree on how that time should be adjusted since they disagree on how much time dilation has occurred. But each observer sees the result of the experiment as entirely consistent with his or her expectations based on the chosen value of ε.

But what if they did the experiment again – this time moving the clock *very slowly* from A to B? After all, the effect of time dilation becomes small as velocity becomes small. So if we move the clocks very slowly, the idea is that time dilation can be neglected. This of course is the idea behind the "slow clock transport" method we described earlier. But this method assumes that time dilation is always negligible in the limit as velocity goes to zero. In fact, this is the case if and only if the speed of light is the same in all directions. But if the one-way speed of light is faster on an outgoing trip than an incoming trip, or vice versa, then time dilation cannot be neglected even at slow speeds. Now that we have the epsilon-dependent time dilation formula, let's see how this works.

Slow Clock Transport Revisited

With the epsilon-dependent time dilation formula, we can see why the slow clock transport method fails to synchronize clocks objectively since it tacitly assumes the one-way speed of light. Namely, we will show that it assumes that $\varepsilon = \frac{1}{2}$ in assuming that time dilation goes to zero. Recall that the slow clock transport method synchronizes two clocks at the same location, and then slowly moves one to location B. The assumption is that the clocks *remain synchronized* in the limit as the velocity goes to zero. But that only happens when $\varepsilon = \frac{1}{2}$. For any other value of ε, only the quadratic term goes to zero in the limit as velocity goes to zero, but a linear term remains, so time dilation still occurs.

Let's consider the motion of the clock that is moved to B, and suppose it is moved a distance of r in time T. The velocity is therefore $v = r/T$. Suppose that we move clock B a distance of one light second (186282.3974 miles). What time does the clock gain or lose relative to a clock that remains stationary? That is, what is $T'-T$ in the limit as velocity goes to zero? Subtracting T from both sides of the epsilon time dilation formula gives:

$$T' - T = T\sqrt{1 - \frac{v^2}{c^2}} \cdot \frac{c}{c + v(2\varepsilon - 1)} - T$$

Since, $v = r/T$, we substitute the value for v in the right-hand term. Then we use the binomial theorem to expand the term under the radical:

$$T' - T = T\left(1 - \frac{1}{2}\frac{v^2}{c^2} + \frac{1}{8}\frac{v^4}{c^4} + \cdots\right) \cdot \frac{c}{c + \frac{r}{T}(2\varepsilon - 1)} - T$$

If we define $X = (r/T)(2\varepsilon\text{-}1)$, then the term following the binomial expansion can be written as:

$$c(c + X)^{-1} = \left(1 + \frac{X}{c}\right)^{-1}$$

This we can also expand using the binomial theorem to:

$$1 - \frac{X}{c} + \cdots$$

Substituting this into the equation, we have:

$$T' - T = T\left(1 - \frac{1}{2}\frac{v^2}{c^2} + \frac{1}{8}\frac{v^4}{c^4} + \cdots\right) \cdot (1 - \frac{r}{cT}(2\varepsilon - 1) + \cdots) - T$$

In the limit as v goes to zero, the v^2 and v^4 terms will also go to zero, and likewise with the higher order terms of X since these are proportional to v^2, leaving:

$$T' - T = T\left(1 - \frac{r}{cT}(2\varepsilon - 1)\right) - T$$

$$T' - T = -\frac{r}{c}(2\varepsilon - 1)$$

Let's first consider the case when $\varepsilon = \frac{1}{2}$. In this case, the term in the parentheses is zero, and we have:

$$T' - T = 0$$

So for slow clock transport, the clocks will indeed remain synchronized, if ε = ½. But for any other value of ε, the result of T'–T will *not* be zero. Suppose ε = 1. In that case the term in the parentheses has a value of one, and we have:

$$T' - T = -\frac{r}{c}$$

In our case, r = 1 light second. So r/c = 1 second. *The clock loses 1 second due to time dilation even in the limit as velocity goes to zero.* Therefore, it is no longer synchronized with a clock that remained stationary at the starting point, even though it was moved as slowly as possible.

We have proved that slow clock transport does not preserve the synchronization of clocks unless the one-way speed of light is the same in both directions (ε = ½). *Thus, the assumption that slow clock transport results in synchronized clocks is logically equivalent to assuming that the one-way speed of light is the same in both directions.*

But then how did Rømer arrive at the correct value for the round-trip speed of light, when he was attempting to measure the one-way speed of light? Essentially, he made two (potentially incorrect) assumptions that exactly cancel. He assumed (1) negligible time dilation and (2) ε = ½ (the one-way speed of light is the same in both directions). These two assumptions are directly related; we saw above that for slow clock transport, time dilation is zero ($T' - T$ =0) only if ε = ½. Of course, if ε = ½ then the round-trip speed of light is the same as the one-way speed.

But suppose that ε = 1, so that incoming light from Jupiter is instantaneous. In that event, Earth's clocks experience time dilation when moving away from Jupiter – even at low velocity. How much time dilation do they experience when they move slowly away from Jupiter by the radius of Earth's orbit? From the last equation above, we saw that when ε = 1, the time lost is: $T' - T = -r/c$. For Earth's orbit, r = 92.9558 million miles, and c = 186,282.3974 miles per second, so $T' - T$ = -8.317 minutes. This, of course, is approximately the time difference that Rømer measured.

Whether we assume ε = ½, ε = 1, or any other value, the observed time difference for Jupiter's moons is always 8.317 minutes over the radius of Earth's orbit. If ε = ½ then this time is the time it takes the light to cross Earth's orbit, and there is virtually no time dilation. If ε = 1 then this is the time difference due to time dilation of Earth's clocks; the time it takes the light to cross Earth's radial distance is zero. For any other value of ε, the result of 8.317 minutes is some combination of genuine time dilation and optical lag due to the time it takes the light to cross the distance.

Time dilation depends on ε in such a way that the *combination* of genuine time dilation and the optical lag due to travel time always sum to the same value regardless of the value of ε. Since it is never possible to empirically distinguish genuine time dilation from optical travel lag, we can never know the value of ε. All experiments that claim to measure the one-way speed of light have tacitly assumed a value for ε in order to achieve (what they think are) synchronized clocks.

All methods by which we might synchronize two clocks separated by a distance tacitly presuppose the one-way speed of light. This is true whether we use fast clock transport, slow clock transport, or signal transmission. Whatever value we assume for ε, the observable results are always the same. We have seen that the value of ε (and hence the value of the one-way speed of light) makes no difference in terms of any objectively measurable results of any conceivable experiment! Whatever value we assume for the one-way speed will result in correct calculations about what happens – as long as we are consistent and use the full ε-dependent derivations for time dilation and length contraction.

I must emphasize that this works only as long as the observer is *consistent* in his or her choice. For example, we may choose the speed of light to be instantaneous when moving directly toward us, and ½c when moving away. That works. But we may not switch (in the middle of a computation) and redefine the one-way speed of light to be instantaneous in some different direction as well. That will not work. The physics requires that we are self-consistent in our choice – whatever that choice is. And we are permitted many choices. We could take the speed of light to be instantaneous when moving north to south, so long as the speed from south to north is ½c. We could take the speed of light to be instantaneous when moving toward John and ½c when moving away – providing John is an inertial reference frame. Amazingly, as long as we are consistent, this choice works too even though the direction of infinite speed depends on where the light is located relative to John.

But why is this so? The time dilation formula depends on ε in such a marvelous way that we can never distinguish the effects of genuine time dilation from the optical lag due to light travel time. It seems as though we are free to choose the one-way speed of light, and the universe confirms our choice. But why? Is there a reason that the math works out this way?

There are three possible propositions available to us concerning the one-way speed of light. The first position is that there could indeed be some experiment to objectively measure the one-way speed of light; we just haven't thought of it yet. The second position is that the one-way speed of light does have an objective value (probably the same in all directions), even though it is impossible *even in principle* to discover what it is. The third position is that the one-way speed of light is not objectively meaningful; it doesn't have any particular value except what we define it to have.

Option 1 is probably the default position for people unfamiliar with relativistic physics. It certainly seems like we should be able to measure a one-way speed by some objective method of synchronizing distant clocks. The fact that many people have proposed experiments that allegedly measure the one-way speed of light suggests that this is a popular position. However, the lack of success of any of those experiments over the past century suggests that option 1 is likely wrong.

Having reviewed the abysmal failure of any experiment – even in concept – to measure the one-way speed of light, our inclination is to move to option 2. Surely the one-way speed of light has a definite value. God knows what it is. But He has constructed the universe in such a way that we can never know what it is. This is certainly a possibility. But as we learn more about the physics of Einstein, this option comes to resemble the idea of the invisible luminiferous ether. Before Einstein, many people believed in an absolute velocity frame – the frame in which the ether is stationary. But since the laws of physics work equally in all inertial frames, we can never discover the frame of the ether. This eventually led to the rejection of an absolute velocity frame. All velocities are relative.

Perhaps the one-way speed of light is analogous to this situation, making option 3 very attractive. Namely, perhaps there is no objective, observer-independent value for the one-way speed of light, just as there is no absolute velocity rest frame. In the next chapter, we will see why the third option is most likely the correct one.

Chapter 18

The Conventionality Thesis and Distant Starlight

By way of review, we have found that the one-way speed of light cannot be measured by any experiment without first assuming the one-way speed of light. To measure any one-way speed requires us to have two clocks separated by some distance that are exactly synchronized. But the only way to synchronize two clocks separated by a distance would require us to know in advance the one-way speed of light. Using radio, light, sound, or other transmissions to set clock B to match clock A does not result in objective synchronization, because all such signals depend on the one-way speed of light.

Furthermore, we cannot achieve synchronization by slow clock transport. To synchronize the clocks at the same location and then slowly separate them does not ensure that they remain synchronized throughout the journey because the act of moving the clocks may cause time dilation. This is true even in the limit as the velocity goes to zero. We know from the epsilon time dilation formula that the amount of time dilation depends on the one-way speed of light – the very thing we do not know in advance.

It is a strange catch-22: we must know the one-way speed of light to synchronize clocks, and we require synchronized clocks to measure the one-way speed of light. *Each goal requires that we achieve the other one first.* Therefore, we cannot objectively establish either one. Consequently, we can never know what the one-way speed of light really is. This suggests something profound about the universe. It suggests that the "correct" value of the one-way speed of light may not be a meaningful idea. Perhaps there is no such thing as an *objective* (observer-independent) one-way speed of light.

By analogy, suppose that Bill and Mary disagree about the length of a table. Bill insists that the table is two units long, whereas Mary claims that it is six units long. The answers are not even close, and each person is perplexed that the other one is so far off. Bill double checks. He gets out his tape measure, places it across the table, and reads off the value: "2." Mary likewise lays her tape measure across the table in the same way, and reads the value: "6." How can this be?

After some deliberation, Mary comes to realize that Bill is using the "wrong" unit of measurement. He is reading the distance in *yards*, where as she measures the distance in *feet*. We might think that this would resolve the disagreement. But suppose Mary said, "See Bill, the reason you were *wrong* about the length of the table is because you were using incorrect units. You were using yards, but of course the *correct* unit of length is feet."

Bill is not at all happy with that answer and responds, "No Mary. Your answer is the wrong one. Clearly the table *really* is 2 yards long. You were using the wrong unit of measurement. You thought the table was six units long because you were using the wrong unit – feet. When the correct unit of length is used (yards), you will see that the table is two units long."

Such a conversation would be surprising to us, because we understand that the unit with which we measure lengths is ultimately arbitrary. It is perfectly fine to use feet, yards, inches, or centimeters. Hence the table is both 6 feet long and also 2 yards long. The numbers are different, but they really correspond to the same objective length. Both Bill and Mary are correct in their measurements, and neither can objectively claim to be using the *correct* unit: there is no such thing.

The one-way speed of light is very much like this scenario. Just as there is no "correct" unit of length, there is no "correct" value of ε. Bill and Mary are each allowed to choose the unit by which they measure lengths. As long as they are consistent, either "feet" or "yards" is a legitimate choice. Likewise, Emily and Jake are each allowed to choose a value for ε, and their choice will determine how they synchronize distant clocks. They will disagree on what constitutes synchronized clocks just as Mary and Bill disagree on the number of units for the length of a table. No experiment can ever establish the "correct" value of the one-way speed of light any more than an experiment can determine the "correct" unit of length.

Some readers may object to this analogy. "That's different! The one-way speed of light is not something we can just choose." But my point is that all theoretical and experimental evidence suggests that indeed the one-way speed of light is something we can freely choose. That may bother us. It may seem to us that there really must be a *correct* objective value for the one-way speed of light, if only we were clever enough to devise an experiment to measure it.

Then again, it may seem to Mary and Bill that there is a correct unit of length. They simply disagree on what it is. But what kind of experiment could we perform to test this claim? We could, of course, take an object whose length is known, and have Mary measure it in feet, and Bill measure it in yards – whoever gets the correct answer is using the correct unit of length.

So the two find an office that is reported to be 12 units in length. Mary and Bill each measure the length. Mary finds that the office is indeed 12 feet in length, but Bill finds that it is only 4 yards in length. Mary wins. Right? But upon further investigation, Bill finds that the previously-established length of 12 units for the office was measured in feet – the *wrong* unit (from his perspective). So naturally, Mary's answer agrees because she too was using the wrong unit. Bill will not be satisfied until they find an object whose length has been reported in the correct unit – yards. But then Mary will complain that yards are not the right unit with which to measure length.

In order to determine which unit of length is correct, we would first need to know how many units long is a given test object. But in order to know how many units long the test object is, we would first have to choose a unit (like feet or yards) so that we can measure the length. But which unit of length is correct is the very thing we are trying to discover. Each goal requires us to accomplish the other one first. It's a catch-22. Does this sound familiar?

Of course, we understand that the debate between Mary and Bill can never be settled because it is meaningless. Both feet and yards are perfectly acceptable units by which to measure a length, and neither is more or less "correct" than the other. This is because units of length are *conventional*. A convention is something that people have chosen and agree to use as a standard by which to measure other things. What we often fail to understand is that the one-way speed of light is also apparently *conventional*. It is not something we can objectively measure. Rather, it is something we define, and then that choice allows us to measure other things – like whether two distant clocks are synchronized.

The Conventionality Thesis

As strange as it may seem, it appears that the one-way speed of light is not a property of the universe, but rather a humanly-stipulated convention. It is something that we are free to choose, and then our choice allows us to have a definition of whether or not two clocks separated by a distance are synchronized (relative to a given observer). This principle is called the *conventionality thesis*. It seems to be a very well-established principle of physics. It has been the subject of considerable discussion in the technical literature throughout the past century. A handful of physicists have tried to argue against the conventionality thesis, but so far none have been able to disprove it. Most physicists agree that the conventionality thesis is true, and this was certainly the position of Albert Einstein as we will see below.

That being the case, we are perfectly free to select a value of ε and use this to synchronize our clocks. Any value between and including 0 and 1 will do. But most physicists prefer to use a value of ½. There are several reasons why this convention is extremely convenient. Not the least of these is that it makes the math easier. Compare the full epsilon time dilation equation in the previous chapter with the simplified version where $\varepsilon = ½$. Setting $\varepsilon = ½$ means that time dilation may be safely neglected at low velocities where $v \ll c$. Also, setting $\varepsilon = ½$ makes the one-way speed of light the same in all directions, which produces a symmetry that is often convenient in solving certain physics problems. For this reason, $\varepsilon = ½$ is the standard choice for physicists and the one that Einstein preferred as well. But he recognized that other choices were equally *legitimate*, regardless of their perceived convenience.

In his primer book on relativity, Einstein briefly discusses the conventionality of the one-way speed of light in his chapter on the relativity of simultaneity. He is considering an experiment in which someone claims that two lightning bolts have struck the ground at the same time at locations A and B. An observer stands exactly halfway between the two lightning bolts at position M. This observer sees the light from both bolts arrive at the same time. But does this prove that they really happened at the same time? It does so only if the light going from A to M travels at the same speed (and thus takes the same time) as the light from B to M. But these are one-way speeds, so in order to measure them we would already need to have a method of synchronizing distant clocks. Here is how Einstein discusses the matter:

...if only I knew that the light by means of which the observer at M perceives the lightning flashes travels along the length A → M with the same velocity as along the path B → M. But an examination of this supposition would only be possible if we already had at our disposal the means of measuring time. It would thus appear as though we were moving here in a logical circle.[31]

Einstein certainly understood the problem. In order to test the claim that the speed of light is the same in both directions, we would need to have synchronized clocks at A and B so that we could measure time there. But to synchronize clocks we would need to know in advance the one-way speed of light. What was Einstein's resolution to the dilemma? He simply chose a value for the one-way speed of light, and that allows this value to *define* what constitutes synchronized clocks for a given observer. And since choosing the speed of light to be the same in both directions makes the math easier, why not select this value? The one-way speed of light is therefore something that we get to stipulate. Einstein states,

That light requires the same time to traverse the path A → M as for the path B → M is in reality neither a *supposition nor a hypothesis* about the physical nature of light, but a *stipulation* which I can make of my own freewill in order to arrive at a definition of simultaneity. [Emphasis is in the original.]

Einstein's statement perfectly summarizes the conventionality thesis. Indeed, the one-way speed of light is a humanly-stipulated convention – not a hypothesis that can be tested. We are freely permitted to choose the one-way speed of light (within certain constraints), and this value will then define what constitutes synchronized clocks.

Why Is the Conventionality Thesis Counterintuitive?

Despite its success, the conventionality thesis is perhaps counterintuitive. Granted, we cannot construct any experiment to objectively measure the one-way speed of light without first assuming it. But we still tend to think that there really is an objective value for the one-way speed of light, if only

---

[31] *Relativity: The Special and General Theory*, fifteenth edition. Authorized translation by Robert W. Lawson. Crown Publishers, Inc. New York, 1961.

God would "whisper in our ear" what it is. We don't have this kind of objection when it comes to feet vs. yards. We understand that we are free to define a unit, and then that tells us how to measure lengths. But why are we reluctant to think that the one-way speed of light is similarly conventional?

One reason might be that we tend to think in terms of absolute time. It seems like the idea of "right now" at some distant location has some objective meaning, and should not depend on the observer's position or state of motion or any particular value of the one-way speed of light. Yet, the real universe just isn't that way. What happens "right now" at a distant location indeed depends on the reference frame of the observer.

We saw that this must be the case from the "trapping the train" paradox. When one observer considers two events to happen at the same time (the closing of the tunnel doors) the other observer considers one event to happen significantly before the other. This is the case even if the two observers are (for an instant) at the same location, but one is moving relative to the other. There is no such thing as objectively (observer-independent) synchronized clocks separated by a distance. We have proved that this is simply the way the universe is. Deal with it.

The relative nature of the flow of time and the absolute nature of the two-way speed of light makes it impossible to have a system of synchronizing clocks that would allow all observers in the universe to arrive at the same answer. By whatever definition we choose for 'synchronized,' and by whatever method we use to accomplish the synchronization of clocks, the clocks will only be synchronized for some observers and not others.

Again, I remind readers that we are not discussing merely what observers "see." Two clocks might be genuinely synchronized, and yet one appears to lag behind the other because it is located at a greater distance, and thus the light takes longer to get to an observer. We are not discussing such optical effects. Rather, we are discussing what happens when observers subtract off any such effects and decide whether the clocks are actually synchronized.

We have come to learn that whether clocks are "actually synchronized" depends on the state of motion of the observer. We learned this from our discussion of the relativity of simultaneity. And now we see that whether clocks are "actually synchronized" *also* depends on the observer's choice for the one-way speed of light.

Objections to the Conventionality Thesis

There are some commonly-used occasional objections to the thesis. Generally, these come not from physicists, but from laymen who have some science background. Though we cannot anticipate every objection, it is instructive to examine a few of the most common.

Perhaps the most common objection is philosophical rather than scientific in nature, and is stated something like this: "Why would the one-way speed of light be different in different directions? Doesn't it make more sense to assume that it is really the same in all directions?" However, this objection doesn't make sense in light of relativity. To measure any one-way speed we would need to use two clocks that everyone agrees are synchronized. But we have already discovered that synchronization is relative to an observer's state of motion. Therefore, there is no objective way of synchronizing two distant clocks such that all observers will agree that they are synchronized. Hence, different observers will disagree on the one-way speed as measured by those clocks. It follows necessarily that one-way speeds cannot be objectively the same for all observers. Thus, one-way speeds are not *objectively* meaningful. They are inherently observer-dependent.

The answer to the objection, then, is to recognize that the question itself is meaningless. The idea that the one-way speed of light is "really" the same in all directions, or "really" different in various directions is like saying that the correct unit of length is "really" feet, not yards. We understand that feet and yards are different ways of measuring length that produce different numbers, e.g. 6 or 2. But neither is any more "real" or "correct" than the other. They are just different. Either one may be used.

A similar objection is that Occam's Razor suggests that the one-way speed of light should be the same in all directions, because this is simpler than the alternative. Occam's Razor is the scientific principle that when two competing models make the same correct predictions, the simpler model is preferred and more likely to correspond to reality. But this objection again fails because Occam's Razor applies to competing *models*, not alternative conventions of measurement. Consider the following:

The metric system is far simpler than the old English system embraced by most Americans. The English system has 12 inches in a foot, 3 feet in a yard, and 1760 yards in a mile. These numbers seem arbitrary. Whereas, the metric system has everything in powers of ten: one hundred centimeters in a meter, one thousand meters in a kilometer. Can we therefore conclude by Occam's Razor that the metric system is more likely to be correct than the English system? It is simpler.

But the English system and the metric system are not competing scientific models. Rather they are different conventions of measurement. They are humanly-stipulated conventions which we use to measure other things. Occam's Razor doesn't apply here since neither the English nor metric system can be said to be right or wrong – only different. Likewise, the one-way speed of light is a humanly-stipulated convention which we use to evaluate or accomplish the synchronization of distant clocks.

Another objection is that Maxwell's equations supposedly show that the one-way speed of light really must be the same in all directions. This requires some explanation. Recall that Maxwell's equations describe the relationship between electric and magnetic fields. These fields can exist even in empty space; electric fields, for example, surround a charged particle. The ability of space to support a magnetic field is referred to as the *permeability* of free space, and is designated by a lower case Greek mu with a zero subscript: μ_0. Likewise, the ability of space to support an electric field is referred to as the *permittivity* of free space, and is indicated by a lowercase epsilon with a zero subscript: ε_0. (Do not confuse this with the one-way speed of light parameter epsilon which *lacks* the subscript.)

One result of Maxwell's equations is the fact that electromagnetic waves in vacuum must propagate at a speed (c) that is related to the permittivity and permeability of free space by the following formula:

$$c = \frac{1}{\sqrt{\varepsilon_0 \mu_0}}$$

Measurements indicate that $\varepsilon_0 = 8.8541878176 \times 10^{-12}$ s^{4}A^2/m^{3}kg and we define $\mu_0 = 4\pi \times 10^{-7}$ kg m/A^{2}s^2. Substituting these values yields $c = 2.9979 \times 10^8$ m/s which is 186282.3974 miles per second. This, of course, is the round-trip speed of light. Yet this solution to Maxwell's equations apparently gives the one-way speed of the light waves as c. Does this disprove the conventionality thesis?

We know by now that there is always a "catch." All four of Maxwell's equations, when expressed in integral form, are closed integrals. This means the act of computing the integral must apply only to a closed surface, *always returning to its starting point.* In a sense, they are all "round-trip" measurements. This suggests that they can only objectively establish the round-trip speed of light, not its one-way speed.

This requires further discussion because the way that Maxwell's equations are usually written *does* imply that the one-way speed of light is *c*. But this is because these equations tacitly assume symmetry, such that ε = ½. Putting it another way, the way that scientists normally write Maxwell's equations is actually a *special case* of a more generalized form; and this special case assumes that the one-way speed of light is the same in all directions, i.e. that ε = ½. So it is a circular argument to claim that Maxell's equations (written in the standard way) prove that the one-way speed of light is the same in all directions since they have been written under that assumption. But they don't have to be written that way.

In the previous chapter we found that John Winnie derived the equations of special relativity without one-way velocity assumptions, leaving ε as a free parameter. We saw that the time dilation and length contraction formulae we derived in earlier chapters were actually special cases of the more generalized *epsilon equations* that Winnie derived. Similarly, in 1978, Dr. Carlo Giannoni derived the general form of Maxwell's equations which leave ε as a free parameter.[32] These general form equations reduce to the standard versions only when ε = ½, but otherwise look a bit different. When we calculate the solution for light waves using the generalized Maxwell equations, we find that the propagation speed of light depends on the direction and value of ε as follows:

$$v = \frac{c}{2\varepsilon} \qquad \text{for negative } x \text{ direction}$$

$$v = \frac{c}{2 - 2\varepsilon} \qquad \text{for positive } x \text{ direction}$$

So when the Maxwell equations are expressed in their full form without assuming ε = ½, then they do allow the one-way speed of light to have a different value in different directions. This is precisely what Giannoni demonstrated in his paper.

Note that Albert Einstein was well aware of Maxwell's equations. Indeed, these formed the basis for his special theory of relativity, in showing that the round-trip speed of light in vacuum is always *c*. Yet Einstein obviously did not believe that these equations establish the one-way speed of light. Recall he stated that the one-way speed of light is "neither a *supposition nor a hypothesis* about the physical nature of light, but a *stipulation* which I can make of my own freewill in order to arrive at

---

[32] Giannoni, C., "Relativistic Mechanics and Electrodynamics without One-Way Velocity Assumptions," *Philosophy of Science*, 45 (1978), pp. 17-46.

a definition of simultaneity." Apparently, Einstein understood that the way Maxwell's equations are normally expressed assumes $\varepsilon = \frac{1}{2}$, and therefore cannot be used to demonstrate that $\varepsilon = \frac{1}{2}$.

The Anisotropic Synchrony Convention

The physics of Einstein requires that the round-trip time-averaged speed of light in vacuum is exactly c. But we have seen that it does not constrain the one-way speed of light which can be anything from $c/2$ to infinity depending on the direction and our choice of ε. Physicists generally prefer to set $\varepsilon = \frac{1}{2}$ because the linear terms drop out of the expression for time dilation and length contraction. This induces a symmetry in the math which makes physics problems easier to solve. This is why we have adopted this convention in the earlier chapters of this book. It is also one reason why Einstein preferred to use this convention. Setting the one-way speed of light to c in all directions is often called the "Einstein synchrony convention" which we shall abbreviate ESC.

The Einstein synchrony convention is also called the "standard synchrony convention," the "Poincaré-Einstein synchrony convention," and the "isotropic synchrony convention." The term 'isotropic' means "the same in all directions." When setting $\varepsilon = \frac{1}{2}$, the one-way speed of light is defined to be the same whether incoming or outgoing. Both are c. But we found earlier that any choice of ε between 0 and 1 will do. Actually, the math works even for values well outside this range, but values below 0 or above 1 do not preserve causality; they have some effects happen before their cause, which violates causality and generally seems unreasonable.

Of the infinite possible choices of ε, two stand out as particularly useful. We already know that $\varepsilon = \frac{1}{2}$ is especially nice because it makes the mathematical equations less cumbersome. Furthermore, this choice imposes a symmetry on spacetime by insisting that the one-way speed of light is isotropic. This means that we don't need to know the direction in which light travels to know its speed. This is very convenient. There is a third advantage to using this convention which we will discuss below.

But the other particularly useful choice is to set $\varepsilon = 1$. Under this convention, outgoing light travels at $c/2$, whereas incoming light travels infinitely fast, reaching its destination instantly. This is an "*anisotropic synchrony convention*" because the one-way speed of light is different for different directions. Technically, all values of ε other than $\frac{1}{2}$ are anisotropic synchrony conventions. But since the $\varepsilon = 1$ choice is uniquely useful, we will refer to it alone as the *anisotropic synchrony convention*.

There is some ambiguity in how we define directions with regard to synchrony conventions. Many physics papers use a rectangular coordinate system (x, y, z), and setting $\varepsilon = 1$ means that light moving in the positive x direction moves at $c/2$, whereas the speed of light moving in the negative x direction is infinite. However, we will use a spherical coordinate system, where $\varepsilon = 1$ means that light moving directly away from an observer moves at $c/2$ (regardless of whether this is in the x, y, or z directions, positive or negative), and incoming light is instant.[33] This is the system adopted by Sarker and Stachel in their paper on the topic.[34]

Why is this convention useful? Under the ASC system all events everywhere in the universe are observed in real time. This is because light travels instantly when directed toward the observer. Therefore, there is no need to subtract off any time lag. It also means that observers do not need to know the distance to the light source in order to compute when an event really happened. Nor do we need to know the distance to a remote clock in order to synchronize our local clock to it.

For example, suppose that on January 3rd, two astronomers (Jake and Emily) observe a supernova – an exploding star. Jake prefers to use the standard synchrony convention (ESC). Emily uses ASC. A few days later, a mutual friend hears about the supernova and calls Jake, asking him when this supernova happened. But Jake has not yet had time to measure the distance to the supernova. So he is unable to answer the question. He passes the phone to Emily. She says, "it happened on the third of January, just a few days ago."

As more time passes, Jake and Emily are finally able to estimate the distance to the supernova; it is 2000 light-years away from Earth, or 1.2×10^{16} miles. Now Jake is able to answer the question of when it happened – about 2000 years ago. Emily maintains her answer that the supernova happened on January 3rd. Both agree that they saw the explosion on the 3rd; the question is, "when did it actually happen?" Using ASC, the light travel time is zero because the light is moving directly toward observers on Earth. It therefore took no time at all to travel the distance of 2000 light-years. Using ESC, the trip took 2000 years.

---

[33] For a given point in space at some distance from the observer, these two different systems can be made to agree locally. Simply take the direction away from the observer and call that positive direction along the x-axis. Other points will have a different x-axis, but it is easy to convent from one point to the next.

[34] Sarkar, S. and J. Stachel, "Did Malament Prove the Non-Conventionality of Simultaneity in the Special Theory of Relativity?" *Philosophy of Science*, 66, June 1999, pp. 208-220.

It is worth noting that from the light's point of view, the trip is instantaneous. This is the case regardless of whether we use ASC or ESC. It is because time dilation becomes infinite in the limit as something approaches the speed of light. So from light's point of view, *every* trip is instantaneous. The question is how long the trip lasted from Earth's point of view. And since this is a one-way journey, the answer is not objectively measurable, but is calculated depending on our choice of synchrony conventions.

It is meaningless to ask whether Jake or Emily has the correct answer to when the event "really" happened. Both are correct according to their chosen convention. This is because the question of *what was the time _on Earth_ at the instant the _distant star_ exploded* is a synchrony question. It requires us to have clocks on Earth synchronized (at least conceptually) with the time of the distant star. The answer will necessarily be observer-dependent. It will depend on the reference frame of the observer, and his or her chosen synchrony convention. The choice is merely a matter of convenience.

Now imagine that the supernova happened 10% slower than expected, suggesting some degree of time dilation. How much time dilation is the star really experiencing? Emily's answer is 10%. In the ASC system, what you see is what you get. But Jake can't immediately answer. He needs to do some math first. If the supernova is indeed time-dilated, then it likely is receding away from Earth. But since it moves away, each light pulse emitted by the star will have to travel a greater distance than the previous pulse, which will take more time. Some of the apparent 10% slowing will be due to genuine time dilation, but the rest will be merely an optical lag due to the increasing distance the light must travel. Emily didn't need to subtract this, because when light is instantaneous, there is no optical lag.

Now suppose that new data arrives from the Hubble Space Telescope, allowing astronomers to make a more accurate estimate of the supernova's distance. They might find that is it actually 3100 light-years away, not 2000 light-years. (These sorts of adjustments happen frequently in astronomy.) What then happens to the estimate of when the event actually occurred? Under ASC, there is no change. Emily maintains that the event happened in the present year on January 3rd. But Jake must revise his estimate since the updated distance means that his previous time lag estimate was wrong. He now claims that the supernova happened 3100 years ago. We again see the advantage of using a distance-independent synchrony convention.

The Relativity of Simultaneity in Various Synchrony Conventions

An interesting consequence of synchrony conventions occurs when we consider the motion of the observers. At the moment Jake and Emily observe the supernova, some astronauts on the International Space Station also happen to observe the event. Let's suppose the space station is in the part of the orbit where it is moving directly toward the supernova, at 7800 meters per second relative to Jake and Emily. The astronauts decide to compute the time that the event happened. Those astronauts who prefer ASC have an easy task. They get the same answer as Emily; the supernova occurred in the present year on January 3rd. But what about the astronauts who prefer ESC?

After estimating the distance to the supernova, the astronauts using ESC compute the time of the event. They find that it happened about 2000 years ago, but they do not get exactly the same answer as Jake. By their estimation, the supernova actually occurred nineteen days earlier than Jake's estimate. This is due to the relativity of simultaneity – see the resolution to the "trapping the train" paradox. Since the astronauts move toward the distant event relative to Jake, they will compute that it actually happened at an earlier time than Jake computes. From the Lorentz transformation from t to t', we can see that the difference is nineteen days.

Even though Jake and the astronauts are both using ESC, they arrive at a different answer for the timing of this distant event. It will be tempting for us to ask who is right. But we understand from relativity that simultaneity, and thus the time-stamping of distant events, is reference-frame dependent. The lack of an objective "actual" Earth time for a distant event is a necessary consequence of physics. But it gets even stranger.

Forty-five minutes later, the International Space Station has completed half an orbit and is therefore on the other side of the Earth. Consequently, it is now moving *away* from the supernova at 7800 meters per second relative to Jake. The astronauts use ESC to again compute the time at which the supernova took place. But now they get a different answer. They now find that the event took place nineteen days *after* Jake's estimate. Again, this follows from the Lorentz transformation for the time coordinate.

What is even stranger is that both estimates are correct. The event took place nineteen days before Jake's estimate and also nineteen days after Jake's estimate, depending on the reference frame. For the astronauts, the correct answer oscillates between these two extremes every forty-five minutes as they orbit the Earth. But those astronauts who use ASC agree with Emily regardless of the direction in which they orbit. They all agree that the event actually happened in the current year on the 3rd of January. Under ESC, the assessment of whether two distant events are simultaneous depends on the observer's velocity. Under ASC, it does not! All observers on Earth would agree on the timing of celestial events if they used ASC – at least to within a fraction of one second. The strangeness of the relativity of simultaneity seems to disappear when using ASC.

Have we eliminated the relativity of simultaneity by using ASC? No, we have merely pushed the strangeness from one place to another as the following experiment illustrates. Suppose two observers separated by a distance of 186,282.3974 miles wish to synchronize their clocks. Both use the ASC system. Dan stands at location **A**, while Marcia stands at **B**. Under ASC, inward-directed light travels at infinite speed. So Dan only needs to read the time at **B** through a telescope, and adjusts the clock at **A** to match. Since the light takes no time to get from **B** to **A**, Dan claims that the clocks are now synchronized.

However, Marcia will not agree. From her point of view, light has infinite speed when moving toward *her*, which means light from **A** to **B** is instantaneous, whereas light from **B** to **A** should take two seconds. Therefore, she will say that Dan's clock at **A** is two seconds behind her clock. According to Dan, the two clocks are synchronized, but according to Marcia they are not. Who is right? Of course, both are right. The idea of synchronized clocks separated by a distance is inherently observer-dependent. There is no one absolute, objective, answer to the question of synchronized distant clocks, any more than there is one absolute, objective answer to feet vs. yards.

There is no synchronization system possible that will allow all inertial observers to agree on whether two distant clocks are synchronized, regardless of the position and velocity of the observers. This is a natural consequence of the physics of Einstein and follows from the way God has chosen to uphold the universe. At best, we can come up with a system that will allow *a subset* of observers with a common property to agree on whether distant clocks are synchronized. But other observers who don't share that property will disagree. And this is where the two natural choices for synchrony conventions come into play.

Under ESC, all inertial observers with the same *velocity* (speed and direction) regardless of their position in the universe will agree on whether two distant clocks are synchronized. But observers with a different velocity will disagree. Conversely, under ASC, all inertial observers with the same position (regardless of velocity) will agree on whether two distant clocks are synchronized. But observers with different positions will disagree – as shown by the thought experiment with Dan and Marcia. So the relativity of simultaneity still occurs under ASC, but in a different way. Under ASC, simultaneity is a function of position, not velocity, whereas under ESC, simultaneity is a function of velocity not position. (All other synchrony conventions will depend on both position *and* velocity.)

Since relativity mainly deals with velocities rather than positions, ESC is a more natural system to use when mathematically computing relativistic effects. This is the third convenience of ESC that I alluded to earlier. The disadvantage of this is that observers on Earth can have very different velocities, and will therefore have different answers to the synchronization of distant clocks. Simply due to Earth's rotation, observers in Australia will have a different exact date for a supernova than those in America. For distant events, the difference can amount to many thousands of years!

On the other hand, since ASC is position-based rather than velocity-based, all observers on Earth will agree on the timing of distant events if they use ASC. This is because we all have basically the same position on cosmic scales. All observers on Earth (and on the International Space Station) are within $1/20^{th}$ of a light-second of each other. Therefore, we will agree to within $1/20^{th}$ of a second on the synchronization of distant clocks, *regardless of our motion*.

In terms of spacetime diagrams, ASC defines the surface of simultaneity as the past light cone of the observer, whereas ESC defines it as the plane horizontal to the time axis that is exactly between the two light cones. Each definition of "now" is inherently observer-dependent, with ESC depending on the observer's velocity, and ASC depending on the observer's position. There is some debate about whether the surface of simultaneity is allowed to be exactly the past (or future) light cone, or merely infinitesimally close. Thus, some papers have the range $0 \leq \varepsilon \leq 1$, whereas other have $0 < \varepsilon < 1$. I maintain the former view, and allow $\varepsilon = 1$, which allows some simultaneous events to be causally connected. But it makes little difference since we may choose ε to be infinitesimally close 1, such as $\varepsilon = 0.99999999999999999999999$ and the difference in synchronization is negligible.

Velocities and Conversions Between ASC and ESC

Under ESC, the one-way speed of light is defined to be c for all directions. For ASC, the one-way speed of light, which we will designate as c_θ is defined by the following:

$$c_\theta = \frac{c}{1 - \cos\theta} \tag{18.1}$$

Here θ is the angle by which the light beam departs from a directly-incoming beam. So for light that is moving directly toward an observer $\theta = 0$, and $c_\theta = \infty$. For light aimed directly away from the observer $\theta = 180°$, and $c_\theta = \tfrac{1}{2}c$. Note that for light aimed at a 90° angle relative to the observer, the one-way speed of light will be c, the same as under ESC. Sometimes critics of ASC will attempt to show that ASC gives the wrong answer for some observation, but they have used the wrong value for the one-way speed, thinking that it is infinite in all directions. It isn't.

It is very easy to convert measurements in ESC to measurements in ASC and vice versa. Let x, y, z, t be the coordinates of an event using ESC. Let $x_\varepsilon, y_\varepsilon, z_\varepsilon, t_\varepsilon$ be the coordinates for the same observer using ASC. Then the conversion between the two will be:

$$x_\varepsilon = x \tag{18.2}$$
$$y_\varepsilon = y$$
$$z_\varepsilon = z$$
$$t_\varepsilon = t + \frac{r}{c}$$

Where

$$r = \sqrt{x^2 + y^2 + z^2}$$

Conversely, we can convert from ASC to ESC. As before, the spatial coordinates remain unchanged. The conversion for the time coordinate is:

$$t = t_\varepsilon - r/c$$

So it is trivial to convert between ESC and ASC. And since the equations of relativity are simplest when expressed in ESC, we have an efficient procedure for evaluating relativistic effects under ASC. Namely, we first convert the coordinates to ESC, do the computation using the simplified $\varepsilon = \frac{1}{2}$ notation, and convert back to ASC. Of course, there is nothing wrong with using the generalized (epsilon) Lorentz transformations to do all computations in ASC directly. But I prefer to do most computations using ESC, just as I prefer to do physics computations using the metric system, since it makes the math easier.

This, in fact, is what all observational astronomers do in practice. All astronomers record the timing of celestial events as they observe them – as if the light had reached the telescope instantly. This is the ASC system. Later, when analyzing the physics of the situation, they will normally convert to ESC. This makes it easier to subtract the effects of Earth's motion since there is no linear term in the time dilation formula under ESC, meaning that time dilation can be safely neglected at low speeds.

From the above equations, we can see that ESC and ASC are merely two different coordinate systems. They each assign spatial and temporal coordinates to any possible event in the universe. And it is possible to convert from one coordinate system to another, just as we can convert from English units to metric and back. Each system has its advantages and disadvantages.

The Distant Starlight Problem

Critics of the Bible often cite distant starlight as a problem for a literal, historical reading of Genesis. The Bible teaches that God created the universe and everything within it in six days and then rested one day (Genesis 1:1-31, Exodus 20:11). That these days of creation are ordinary Earth rotation days is obvious from the context; each day is defined as bounded by one evening and one morning (Genesis 1:5, 8, 13, etc.) and forms the basis of our work week (Exodus 20:8-11). Furthermore, human beings were created on the sixth day (Genesis 1:26-31). There are less than eighty generations between Adam and the birth of Jesus Christ. This implies that the time between Adam and Christ's earthly ministry was a few thousand years. Detailed calculations from the biblical information put the number somewhere around 4000 years. And since Christ's birth was just over 2000 years ago, we must conclude that the age of the universe is around 6000 years.

This sharply clashes with the secular belief that the universe is 13.8 billion years old. The critics claim that distant starlight supports their position, and disallows the possibility of 'recent' (6000 years ago) creation. Namely, there are galaxies that are billions of light-years away. Since light travels at speed c (which is a distance of one light-year in a time of one year), it follows that starlight from these galaxies should take billions of years to reach Earth. There is no doubt that the starlight has already reached Earth, because we can see these galaxies. Therefore, the secularists claim that our ability to see light from distant galaxies is evidence that the universe is billions of years old, and Genesis is wrong.

But when we understand the physics of Einstein, we immediately see some problems with the critics' argument. They are claiming that the time between two distant events (the light leaving the galaxy, and then arriving on Earth) is billions of years. However, they have failed to specify a reference frame, they have ignored time dilation, and most significantly, they have assumed the Einstein synchrony convention. Let's explore just the last of these assumptions.

The light's journey from a distant galaxy to Earth is a one-way trip. Consequently, the time this journey takes will be the distance divided by the one-way speed of light. But of course, the one-way speed of light is conventional. Using ASC coordinates, incoming light has infinite speed. Therefore, the time needed for light to travel from the distant galaxies to Earth is zero. Recall that under ASC, we are seeing the universe in real time. Consequently, our ability to see distant galaxies is perfectly compatible with the Genesis account. There is no distant starlight problem.

The critic might claim that this answer is predicated on the conventionality thesis. But it really isn't. The critic's argument against the biblical timescale is predicated on the assumption that the one-way speed of light must be c. But this certainly has not been demonstrated. Even if the conventionality thesis were refuted, and it was shown that the universe indeed prefers a particular value for the one-way speed of light, that value may well turn out to be infinity for inward-directed photons.

I am convinced that the conventionality thesis is true. But my point is that the solution to the distant starlight issue does not depend on this. In order for the critic's claim (that distant starlight proves a multi-billion-year-old universe) to have any weight, he must first show that the conventionality thesis is wrong, and then also show by experiment or non-circular reasoning that the one-way speed of light is indeed c. And so far, no one has been able to do either of these things. Therefore, at the present time, distant starlight cannot be considered a legitimate or rational objection to the biblical timescale. The critic would have to prove that the one-way speed of light must be *c*, and no one has been able to do this.

Does the Bible Use a Synchrony Convention?

For the aforementioned reasons, distant starlight is not a legitimate objection to recent creation, since the critic cannot prove that ASC is false. Nonetheless, a failure to prove ASC false doesn't make ASC legitimate. As an apologetic defense, we are done. But our curiosity may drive us to wonder if ASC really is the way that light was able to reach Earth in the biblical timescale, or if there is some other answer.

After all, even if ASC is a perfectly valid convention, the Lord is not obligated to use it in Scripture. He might use ESC or some other system. The Bible does occasionally mention celestial events taking place at a particular time; hence, *some* synchrony convention is in place. If the Bible uses ESC as the system by which clocks are synchronized and the creation of the universe took six Earth rotation days by ESC reckoning, then the question of how we are able to see distant galaxies remains unanswered. However, I suggest that the Bible in fact uses the ASC convention for the following reasons.

First, until modern times, ASC has always been the historical standard synchrony convention. Ancient astronomers recorded only the ASC time of any celestial event; we have many of their records today. They did not subtract light travel time. They didn't know the speed of light (either round-trip or one-way). And they didn't know the distance to celestial objects.[35] So it would have been impossible for them to subtract an unknown travel time for the light.

There is no evidence that anyone used ESC before the late 1600s. As far as we know, Ole Rømer was the first person to subtract his own estimate of light travel time from a distant source (the moons of Jupiter). By attributing the advancement and retardation of eclipses of Io to light travel time instead of time dilation, Rømer effectively assumed that the one-way speed of light is the same as its round-trip speed, which is true only in ESC. Before his measurements of Jupiter's moons, no one could

---

[35] The one exception is the moon. In the late 3rd century BC, Aristarchus was able to estimate the distance to the moon (very roughly). He also attempted to measure the distance to the sun, but this latter estimate was not very accurate due to the limitations of ancient measurements. In any case, he did not know the speed of light, and therefore necessarily used the ASC system of time-stamping.

have subtracted the light travel time because no one knew what the speed of light was. Hence, it was impossible for anyone to use ESC before the late 1670s. Since the Scriptures were written long before this date, it seems reasonable that they too would use ASC.

But someone may object by pointing out that the Bible is inspired by God, and God could certainly use ESC long before people had thought of it. Quite right. But *would* God use a convention that would not be implemented by human beings until recent times? If God used ESC to describe the timing of celestial events, but all ancient people groups used ASC, then they would have misunderstood God's Word for nearly 6000 years. They would have understood the words differently from God's intention.

All word usage is inherently conventional. There is no particular reason why the word 'lion' must refer to the large cat we all imagine when we read the word. We simply agree by common consent to define the word as such. If God were to use conventions in the Bible that people had not yet considered, then people at that time would not have correctly understood the Bible. But if that is so, then there is no guarantee that today we have finally stumbled on the correct meaning that God intended. Perhaps that discovery lies a hundred years in our future, or a thousand years! Indeed, if God is willing to use linguistic conventions that are ahead of the time of the people He writes to, then we cannot be sure that we correctly understand anything whatsoever in the Bible.

If that were the case, then we might as well not even bother to read the Bible if God does not use terms that we in the present can understand. But that would be absurd. It is a biblical principle that God is clear and understandable in his Words. God is not the author of confusion (1 Corinthians 14:33).

If we can communicate with and understand each other (most of the time), how much more is God able to clearly communicate with His creations. Thus, we expect that God uses the linguistic conventions of the time and people-group to whom the biblical text was written. This logically extends to the way we speak of celestial events. And we know from other historical documents that the original audience used ASC.

Furthermore, there is internal evidence that the Bible uses ASC. For events taking place on Earth, there is very little difference between ESC and ASC. The two conventions give the same number to within a fraction of a second. But for celestial events the difference is quite evident. Events that are observed in space today actually occurred in the distant past by ESC reckoning. But they take place today by ASC reckoning. Conversely, distant events that take place today are seen today by ASC reckoning, but will not be seen until the distant future by ESC reckoning. So when the Bible touches on celestial events, does it indicate that their light took a long time to reach Earth (ESC), or no time at all to reach Earth (ASC)?

Genesis 1:14-15 states, "Then God said, 'Let there be lights in the expanse of the heavens to separate the day from the night, and let them be for signs and for seasons and for days and years; and let them be for lights in the expanse of the heavens to give light on the Earth'; and it was so." We learn from verse 16 that these lights include the sun, the moon, and the stars. And verse 19 indicates that these were all created on the fourth day. Furthermore, verse 15 indicates that the purpose of these luminaries was "to give light on the Earth." But verse 15 also states, "and it was so." That is, the stars began giving light upon the Earth on the very day they were created. Hence, there is no apparent time between when the stars are created, and when their light reaches Earth. God made the stars on day four and their light reached Earth day four. This is exactly what ASC requires.

Indeed, in all celestial events mentioned in Scripture, there is no hint of any time lag between when the event happens, and when it is observed on the Earth. It is as if the incoming light arrives instantly. Every Scriptural indication seems to describe the universe as if we are seeing it in real time. The creation of the stars seems to be simultaneous with their light reaching Earth (Psalm 33:9, Genesis 1:14-15).

There cannot be a distant starlight problem under ASC since the universe appears in real time. What critics take to be a problem is in reality an indication that the Bible uses ASC, consistent with all other ancient literature. To object to the Bible's use of ASC makes about as much sense as criticizing it for using cubits rather than the metric system. Nonetheless, some people have suggested that ASC is not legitimate. We will explore these objections in the following chapter.

Chapter 19

Objections to ASC

Several years ago, I published an article on the one-way speed of light, arguing that the Bible likely uses ASC when describing the timing of events. The article appears in the peer-reviewed creation science journal: *Answers Research Journal*. It is free and available here:

https://answersingenesis.org/astronomy/starlight/anisotropic-synchrony-convention-distant-starlight-problem/

For readers interested in this topic, I highly recommend reading this paper. It covers some of what we have discussed from a somewhat different perspective and makes some additional points that I will not repeat here. The paper also presents a preliminary cosmological model based on the premise that God created the universe in six days by ASC timing. The paper passed peer review and was published in 2010. Since then there have been no rebuttals or criticisms in any peer-reviewed technical literature aside from a passing remark in one CRSQ article that I will discuss below.

But there have been a number of internet posts that claim to refute ASC – either the convention itself or the cosmological model that I developed on the basis that the Bible uses ASC. None of these are logically cogent counter-arguments, which is presumably why they have not appeared in technical science journals; they would never pass peer review. This is why scientists do not take internet posts seriously, and neither should you. Anyone can post an opinion on the internet, but that does not mean that the post has any rational or scientific merit whatsoever.

Nonetheless, it may be instructive to review a few of these objections to see why scientists do not take them seriously.

A Gravitational Field?

We start with the claim that using the ASC system would create an enormous gravitational field centered on Earth (in addition to Earth's gravitational field). No such field is observed. Hence, the critic concludes that ASC is wrong.

In at least one case, the critic seems to agree that the conventionality thesis is correct – that the one-way speed of light is indeed conventional, with only the round-trip speed being experimentally testable. Hence, the speed of light can be up to infinite in any one direction. But he or she suggests that the direction in which light is infinite must be the same throughout the universe as measured in rectangular coordinates.

In other words, the critic alleges that indeed the speed of light may be infinite when traveling north to south; but then for every point in the universe the infinite direction must also be north to south. Conversely, ASC uses a spherical coordinate system which has the infinite-speed direction as "toward the observer." That direction will be north-to-south for light that is north of the observer, and south-to-north for light that is south of the observer. But, does the universe really allow only rectangular coordinate systems?

This critic claims that to have the speed of light be infinite in the direction of the observer (wherever he or she is) is disallowed. The claim is that such a coordinate system would create a gravitational field, although the details of the critic's reasoning are shaky. In one version of this criticism, the critic seems to be bothered that the velocity of the one-way speed of light can change as it travels: for example, if the beam is at some arbitrary angle so that it misses the observer. The light would be faster upon approach, and slower upon departure. From what I can tell, the critic has made the false assumption that changes to the one-way speed of light are always due to a gravitational field, rather than the generalized Maxwell equations.

The critic's mistake is the logical fallacy of the "commutation of conditionals." For any two propositions p, and q, the fallacy is as follows:

1. If p then q.
2. Therefore, if q then p.

One example might be, "If it's raining outside, then the grass will be wet. Therefore, if the grass is wet, then it must be raining outside." This fails because there are other causes of wet grass – sprinklers for example. It is true that "if there is a gravitational field, then the velocity of light will be affected." But this does not imply the reverse – that a change in the velocity of light requires a gravitational field. There are other ways in which the one-way speed of light might change. Indeed, the speed of light changes dramatically when photons enter a material like diamond. *But this does not create a gravitational field!*

In another version, the critic seems to be assuming an ESC-based definition of the spacetime interval, and then arguing that this proves ESC, or at least a one-way speed of light constant along rectangular coordinates.[36] But that would be circular reasoning. One may not assume ESC for the sake of proving it. We have already noted that many equations take on a slightly different form when $\varepsilon \neq \frac{1}{2}$, such as the generalized (epsilon) Lorentz transformations we saw in the previous chapter.

Furthermore, these critics seem to be unaware that the spherical coordinate system embraced by ASC is *already established* in the secular peer-reviewed technical literature. As one example, Sahotra Sarker and John Stachel used the past light-cone as the surface of simultaneity in their research paper – they took the incoming one-way speed of light to be infinite with respect to a given observer.[37] They discussed some of the advantages of this system. And they certainly did not do this to solve any perceived distant starlight problem. Sarkar is adamantly anti-creationist.

Finally, it is very easy to experimentally prove that ASC coordinates do not create a measurable gravitational field. Simply buy a clock that synchronizes itself to the radio transmissions from the station in Fort Collins, Colorado, which is linked to the atomic clock. All such clocks use ASC because they do not subtract any light/radio travel time. If you turn on the clock, and Earth collapses into a black hole, then the critic is right and using ASC creates a gravitational field. But, of course, that will not happen. I have such a clock, and it has been working well for years using ASC, with no induced gravitational field.

---

[36] We will not give the proof here, but the square of the spacetime interval leaving ε as a free parameter is:
$$ds^2 = dx^2 + dy^2 + dz^2 - c^2 dt^2 - (2\varepsilon - 1)^2 dx^2 + 2c(2\varepsilon - 1)dxdt$$
[37] Sarker, S, and J. Stachel, "Did Malament Prove the Non-Conventionality of Simulteneity in the Special Theory of Relativity?," *Philosophy of Science* 66 (1999), pp. 208-220.

Switching coordinate systems cannot create any real forces, because coordinates are merely our way of cataloging events. When we understand that ASC is a coordinate system, marking the x, y, z, t of any event, we realize just how utterly ridiculous the critic's claim really is. How can switching to a different measurement system possibly induce a gravitational field?   What if we applied that reasoning to other measurement systems? Suppose someone said, "You must use inches to measure a distance; if you use centimeters it will create an enormous gravitational field and the Earth will be sucked into a black hole!" Would that be reasonable? If not, then neither is it reasonable to apply the same argument to ASC.

So the claim that ASC would create a gravitational field may sound intimidating to those unfamiliar with physics, but it is really quite silly. Yet, many critics have jumped on this. Even Hugh Ross's ministry "Reasons to Believe" seems to have been fooled by this objection.[38] But the system we choose to catalog various aspects of the universe has no effect on the actual universe.

Maxwell's Equations

Another common objection is that Maxwell's equations disprove ASC because they show that the one-way speed of light is isotropic. But we saw in the previous chapter that the full, generalized Maxwell equations do nothing of the kind. They only predict isotropic waves in all directions when $\varepsilon = \frac{1}{2}$. Under any other value of ε, they predict non-isotropic one-way wave velocities. The generalized version of Maxwell's equations allows for non-isotropic light propagation in cases where $\varepsilon \neq \frac{1}{2}$. In ASC, $\varepsilon = 1$; therefore, non-isotropic light propagation is permitted by Maxwell's equations under ASC.

I will not repeat the details here since we covered this in the previous chapter. But I will add that there are certain materials, such as calcite crystals, where the speed of light can be quite different depending on the direction or polarization axis of the light. Yet, no physicist would argue that Maxwell's equations disallow such phenomena. Maxwell's

---

[38] http://www.reasons.org/articles/an-infinite-speed-of-light <accessed 4/11/2016> Note that this article seems to have been pulled. Perhaps the Reasons to Believe staff realized that it is fallacious and removed it? But there has been no public retraction. How many people were misled by this article, without correction?

equations continue to apply even in non-isotropic media. So to object to ASC on this basis is quite uninformed.

Light Echoes

Some critics have suggested that "light echoes" disprove ASC. A light echo is the result of light reflecting off a nebula in which we observe the reflection at a later time than we observe the source. For example, a star will temporarily brighten. Some of its light travels directly to Earth and we observe the brightening. Some of its light travels at a 90-degree angle from Earth. Then the star fades. But the light traveling at 90 degrees strikes a nebula nearby the star, and some of it reflects toward Earth. So we see the nebula brighten at a later time than we saw the star brighten.

Doesn't this mean that the light took some time to get from the star to the nebula, and thus its speed is not infinite? Yes, it does. And that is exactly what ASC predicts. Recall that ASC defines the one-way speed of light that is *directly toward an observer* as moving infinitely fast. Light that is at 90 degrees from the observer's line of sight will move at speed c: the same as the round-trip speed. And light moving directly away travels at $\frac{1}{2}c$. Critics who make this claim have misrepresented ASC, as though it has the one-way speed of light infinite in all directions. It doesn't. Light echoes are predicted under both ESC and ASC, and the observed timing is identical.

Redshifts and the Doppler Effect

Another objection involves redshifts. Why do we see a Doppler effect if inward-directed light is instantaneous? Recall from an earlier chapter that the Doppler effect is due to the combined effects of (1) the increasing or decreasing time lag of the light as it has to traverse a changing distance, and (2) time dilation. Under ASC, for incoming light there is no time lag since inward-directed light is instantaneous. But there is still an effect of time dilation, and the equation is a bit different from the ESC version. So there will still be a redshift due to this time dilation. But will it be the same redshift predicted by ESC?

Suppose a light source moves away from an observer at speed v. As we discovered in chapter 11, according to ESC the redshift is given by equation 11.3:

$$\frac{\lambda_O}{\lambda_S} = \sqrt{\frac{1 + \frac{v}{c}}{1 - \frac{v}{c}}}$$

What will the redshift be according to ASC? Since the light travels directly toward the observer with infinite velocity, there is no time lag in ASC. So any redshift will be due entirely to time dilation. Therefore, it follows that:

$$\frac{\lambda_O}{\lambda_S} = \frac{T}{T'}$$

Here T is the time between two ticks of the moving clock as measured in the stationary frame, and T' is the time measured in the moving frame. But the time dilation formula for ASC is different from the ESC version. Since $\varepsilon = 1$ for ASC, from equation 17.1 the time dilation at velocity v is:

$$T' = T\sqrt{1 - \frac{v^2}{c^2}} \cdot \frac{c}{c + v}$$

With some algebra, this reduces to:

$$\frac{T}{T'} = \frac{1 + \frac{v}{c}}{\sqrt{1 - \frac{v^2}{c^2}}}$$

Substituting the wavelength ratio for the time ratio and with further algebra we have:

$$\frac{\lambda_O}{\lambda_S} = \frac{T}{T'} = \frac{1 + \frac{v}{c}}{\sqrt{\left(1 + \frac{v}{c}\right)\left(1 - \frac{v}{c}\right)}} = \frac{\sqrt{1 + \frac{v}{c}}\sqrt{1 + \frac{v}{c}}}{\sqrt{\left(1 + \frac{v}{c}\right)\left(1 - \frac{v}{c}\right)}}$$

This reduces to:

$$\frac{\lambda_O}{\lambda_S} = \sqrt{\frac{1 + \frac{v}{c}}{1 - \frac{v}{c}}}$$

So the formula for redshift is identical for ASC and ESC. What was observed due to a combination of time dilation and optical time lag under ESC is entirely time dilation under ASC. It is impossible to distinguish *observationally* the effect of genuine time dilation from the effect of increasing time lag of the light. We only ever observe the combination. The fraction of the redshift due to time lag under ESC is absorbed into the increased time dilation under ASC.

The galactic redshifts due to expansion of the universe are likewise identical between ASC and ESC. What was due to increasing distance in ESC is due to time dilation in ASC. But the observed result is always identical. So redshifts are not a problem for ASC; they are predicted to be identical to redshifts under ESC.

None of the above objections have any scientific merit. They are based on a misunderstanding of physics. Hence, they would never be published in a quality peer-reviewed science journal because any reviewers trained in physics would have spotted the obvious blunders. This is why we don't spend a lot of time responding to internet claims. Most of them are quite silly.

Do Binary Stars Establish the One-Way Speed of Light?

But one criticism of ASC did appear as a brief comment in the *Creation Research Society Quarterly* (CRSQ) in a paper that was about a different topic. The paper was written by a friend of mine, Dr. Ron Samec, who is an observational astronomer. He is an excellent researcher, and most of the paper is accurate. But Samec did take a paragraph in the conclusion to argue that his observations disqualify the ASC model that I had proposed. He later commented that this was mainly to stimulate discussion on the topic, which I appreciate.

Some people are uncomfortable when one scientist challenges another; they sometimes take it personally. But in fact, good scientists appreciate criticism. It helps us to catch errors, and to think through issues. After all, we want to be correct in our conclusions. I do believe that Samec's remarks about ASC are mistaken because he neglected to account for time dilation in the moving source, as I will demonstrate below. But I appreciate his challenge, and I reply in the same iron-sharpening-iron spirit. Basically, Dr. Samec's observations involved the timing of eclipses in a binary star system. It is similar in principle to Rømer's observations of Jupiter's moons. Samec's observations fail to establish the one-way speed of light for the same reason as Rømer's, but they do establish the round-trip speed.

Before I explain the error and its resolution, I must point out that Dr. Samec did one thing right: he published in peer-reviewed literature. The peer reviewers apparently did not catch the error, possibly because the conventionality thesis is not well known even among many physicists. In any case, since Samec published in peer-reviewed literature, this gave me an opportunity to respond in the same literature. In my reply, I pointed out that Samec had neglected relativistic time dilation. In ASC, time dilation has a quasi-linear relationship to velocity, and thus it may not be neglected even at slow speeds. I further pointed to John Winnie's paper and referenced the specific equation that gives the full time dilation formula with epsilon dependence.

Specifically, I pointed out that "the relativistic time dilation term has a linear component in the ASC model which cannot be neglected even at relatively low velocities." In neglecting time dilation, therefore, Samec had tacitly assumed ESC. He had assumed that the advancement and retardation of observed eclipses was due entirely to the changing distance that the light must traverse at finite speed. But he neglected to compute the effect of time dilation which also contributes.

Samec responded by asking, "How could these be due to just relativistic effects?" But I had *already* answered that question in my original reply. Namely, "the relativistic time dilation term has a linear component in the ASC model which cannot be neglected even at relatively low velocities." I suspect that Samec did not read Winnie's paper, or the equation I referenced, because this would have then been clear. Samec then stated in his reply that the advancement and retardation effects of observed eclipse timings "are due to the light traveling across the system at the known speed, c." But of course, this *begs the question*. Only under ESC would this be true. Under ASC, the effect is due entirely to time dilation. Below I will explain specifically how this works.

The observations involved a binary star system FY Bootis. This is a three-star system. Two of these stars are contact binaries; they orbit around their common center of mass so closely that they essentially touch each other. The close orbit results in a very small orbital period: less than six hours! Their orbital plane is roughly aligned with Earth such that these two stars eclipse each other from our perspective. By observing the timing of the eclipses, we know the orbital period very precisely.

A third unobserved star orbits at a greater distance. Its existence is inferred by the gravitational influence on the observed binary pair. Since all three stars orbit the center of mass of the system, the binary pair is closer to Earth when the third star is farther, and is farther from Earth when the third star is closer. Furthermore, we observe that the observed eclipse timings happen a bit earlier when the binary pair is closer to Earth, and later when the binary pair is farther from Earth. Samec interprets this as due to the changing distance that light must travel at (one-way) speed c. However, under ASC, it is due to time dilation. But do both ESC and ASC give the same answer?

Let's see how this works mathematically:

From Samec's data and other observations, the average orbital period of the binary pair is 0.24115955 days, $\pm$0.00000005 days. The orbital period of the third companion is 9.9 years $\pm$0.2 years. Based on the equation listed in Samec's Figure 8, the difference in eclipse times between the binary's closest and farthest distance from Earth is 0.005124 days, which is 7.379 minutes. Samec attributes this difference to light travel time (he's assuming ESC) which works out to a distance of 8.247×10^7 miles for the radial component[39] of the third star's orbit. So when the binary stars are at their farthest point from Earth, the observed eclipses are delayed by 7.379 minutes from their time a half orbit earlier. Under ESC, the time dilation is negligible, and the 7.379-minute difference is due to the time it takes the light to travel (one-way) across the orbit.

Under ASC, the time it takes for Earth-bound light to cross the system is zero, and the time difference is due entirely to time dilation. But does it give the same result as ESC? Using the epsilon time dilation formula for ASC (17.1) in which $\varepsilon = 1$, we have:

$$T' = T \sqrt{1 - \frac{v^2}{c^2} \cdot \frac{c}{c + v}}$$

---

[39] The true radius of the orbit will be somewhat larger because the system is tilted with respect to Earth. Only the radial component of motion (that which is Earth-directed) affects the distance and thus contributes to the time difference.

Since the velocity of the binary pair is very small compared to the speed of light, v^2/c^2 will be infinitesimal compared to the other terms. Therefore, we may safely neglect this term, and the equation simplifies to:

$$T' = \frac{Tc}{c + v}$$

The average velocity of the binary star is simply the distance it travels divided by the time it takes: $v = s/T$. Here s is the total distance of 8.247×10^7 miles and T is half of the orbital period of the third star (since half an orbit moves the binary stars from their closest point to their farthest point from Earth). So $T = 4.95$ years $= 1.562 \times 10^8$ seconds.

$$T' = \frac{Tc}{c + \dfrac{s}{T}}$$

The time lost by the binary due to time dilation will be $T-T'$. It follows that:

$$T - T' = T\left(1 - \frac{c}{c + \dfrac{s}{T}}\right)$$

All the values on the right-hand side of the equation are known. Substituting these values, we find:

$$\boxed{T - T' = 7.379 \text{ minutes}}$$

So the ASC model predicts that the binary stars will experience a loss of 7.379 minutes due to time dilation as they move from their closest point in the orbit to their farthest point. This is of course exactly what Samec observed. Therefore, *his observations are perfectly consistent with ASC.* They are equally consistent with ESC of course. And this must always be the case since ESC and ASC are merely coordinate systems. They cannot differ in their prediction of any observable phenomenon.

When we really understand that ASC and ESC are merely two different coordinate systems, we can see why it is meaningless to talk about which is "true." Each is a legitimate system by which to catalog events. Under ASC, we see the universe in real time. And since this seems to be the convention used in the Bible, there can be no distant starlight problem for a young universe. There never has been.

Chapter 20

Conclusions

The physics of Einstein is overwhelming in its implications, and yet elegantly simple in its starting axioms. Merely by presupposing the relativity principle, and the constancy of the round-trip speed of light in vacuum, we can mathematically prove that time dilation, length contraction, and mass increase, must occur. Scientists have confirmed these effects. But even if they hadn't, we could be assured that such effects are real on the basis of pure deductive reasoning. Very few fields of science are developed in such a way.

Relativity challenges our intuition. We tend to think of time, distance, and mass as "absolute" quantities that are unaffected by the observer's velocity. This misconception is quite natural since relativistic effects are utterly miniscule at ordinary speeds. Only when velocities reach a substantial fraction of the speed of light do effects like time dilation become noticeable. We simply have no practical experience traveling at such velocities. We tend to infer that velocity does not affect that passage of time or lengths or masses because such effects are ordinarily far too miniscule to be observed.

In principle, time dilation would allow a fast-moving astronaut to travel to a distant star and back, while aging very little. He would essentially be time-traveling into the future, and could potentially meet his distant descendants. From his point of view, the trip took little time due to length contraction of the universe. This twin paradox may be counterintuitive, but it is perfectly self-consistent and provable on the basis of known physics. However, time travel into the past seems to be forbidden. A person would have to travel faster than light; and no amount of energy can accelerate ordinary matter up to or past this ultimate speed limit. Additionally, the law of non-contradiction would apparently disallow any information-bearing system to travel backward in time.

As if these effects weren't strange enough, even the idea of simultaneous is relative. There can be no such thing as a universal "now." Two observers at different positions and velocities will not generally agree on whether any two distant events are simultaneous. Furthermore, there is no system of synchronization that will bring all observers into agreement. Therefore, if two clocks are separated by some distance, there can be no objective answer to the question, "Are these two clocks synchronized?" The relativity of simultaneity is derivable from first principles, and is essential to understanding the resolution to the "trapping the train" paradox.

The relativity of simultaneity is not merely an optical effect, but persists even when each observer has subtracted any perceived light travel time. The speed of light divides all of spacetime into two regions – timelike regions within the light cones, and spacelike regions outside of them. If the two events are spacelike, the observers might not even agree on which event happened first. Our temptation is to think that one of the observers must be wrong. But the impossibility of an objective universal "now" indicates that the order of spacelike events is non-objective, and depends on the reference frame.

The physics of Einstein reverses the apparent roles of light and spacetime. Our intuition expects that space, time, mass, and simultaneity are absolute, and all speeds are relative to the observer. In reality, the round-trip speed of light is absolute, and space, time, mass, and simultaneity are all relative to the observer. Rather than light submitting to the universe, it is as if the rest of the universe submits to light, bending and stretching in just such a way as to keep light constant. How delightfully unexpected this is!

Mass (relativistic mass) is also reference-frame dependent. As we increase the speed of an object, we are adding energy to it. But that energy actually has mass, making the moving object "heavier" and therefore harder to accelerate. To accelerate a mass up to the speed of light would take an infinite amount of energy, and therefore cannot be done. But massless particles (those with no rest mass) *must* travel at the speed of light if they have any energy at all. The particles comprising light (photons) therefore have no rest mass. Of course, they are never at rest and they do possess relativistic mass.

Experiments have confirmed that time dilation, length contraction, and the increase of relativistic mass do occur at precisely the values we derived from deductive reasoning from first principles. Amazingly, these equations can also be applied with elegant consistency to situations that apparently never occur in the universe. For example, we know how tachyons would behave if they existed, even though they apparently do not exist. We know that tachyons would behave in a "backward" fashion, speeding up as they lose energy, and slowing when energy is added. They can never quite be slowed to the speed of light; just as ordinary matter can never be accelerated up to the speed of light.

Further, tachyons would be spacelike; their arrival would always be outside the light cone of their departure. And since the order of spacelike events is dependent on reference frame, tachyons would necessarily be traveling backward in time for some observers. That is, the tachyon would reach its destination before its departure. Since spacelike events are non-causal, tachyons could not "affect" anything. This leads us to suspect that they do not exist in reality, even though we understand the properties they would have to possess.

We have seen that the coordinate system we use to label the time and location of distant events is ultimately conventional – a matter of human choice. And since coordinate systems are required to measure any one-way velocity, the one-way speed of light (as well as any one-way velocity) is inherently conventional. It is something we may select within certain constraints, as long as the round-trip speed of light remains c. As such, arguments that the universe must be very ancient on the basis that we can see distant starlight, have no merit. In relativity, the speed of light in any one direction may be defined to be as high as infinity. However, the speed of light in the opposite direction must always be such that the time-average is always c in vacuum.

The Reasonable Effectiveness of Mathematics in the Natural Sciences

Lorentz aberration and beaming, relativistic redshift, gravitational time dilation, and black holes are all derivable from the basic principles of relativity. The physical universe does correspond to the predictions of the physics of Einstein. And yet this physics was primarily discovered not by observations of the universe, but by rigorous application of logic and mathematics: mental exercises. Thus, we knew that time dilation must occur long before it was experimentally confirmed. We know how the inside of a black hole works, even though no one has ever been there. How is it that the human mind is able to discover truths that go far beyond anything we've experienced?

The biblical worldview provides us with the answer. God upholds the universe in a consistent and orderly fashion and has promised that He will continue to do so (Hebrews 1:3, Genesis 8:22). Logic is the study of the principles of correct reasoning. In the biblical worldview, God is the source of all truth, and therefore laws of logic are essentially a reflection of the way God thinks. Mathematics is the way God thinks about numbers. And since it is the mind of God that upholds the physical universe, naturally the physical universe will be characterized by strict adherence to logical and mathematical principles.

Furthermore, God has made humanity in His own image, after His likeness (Genesis 1:26-27). This does not mean that we physically resemble God; God is an immaterial, invisible Spirit (John 4:24, 1 Timothy 1:17). Rather, human beings reflect in a finite way some of God's attributes. Among these is the capacity for rational thought. The Lord granted human beings with minds that are capable of thinking in a way that is consistent with the mind of God (Isaiah 55:7-8). So we can know about the inner workings of a black hole not by experience, but by application of God's laws of logic and mathematics, knowing that God upholds all things by His power.

We can have confidence that the laws of physics will apply in the future just as they have in the past because God is beyond time and laws of physics reflect the consistent way He upholds nature. We can be sure the laws of physics will apply in all locations in the universe, because God is sovereign over all of nature. We can have confidence that the physics of relativity that we have discovered almost entirely by logical deduction from first principles will apply to the physical universe because the God of logic is also the God of nature.

But the secular worldview cannot account for the success of the physics of Einstein, or any other science for that matter. Without God, how can we possibly know that laws of logic apply everywhere and at all times? How do we know that mathematics works the same inside a black hole as it does on Earth? How could we ever know how tachyons would behave, since they apparently do not and cannot exist?

In 1960, the Nobel-Prize-winning physicist Eugene Wigner published an insightful paper entitled "The Unreasonable Effectiveness of Mathematics in the Natural Sciences." Writing from a secular perspective, Wigner pointed out that there is no basis for expecting that the physical universe should obey mathematical laws. After all, mathematics is conceptual – it exists in the mind. But the physical universe is material – it exists outside the mind. How can discoveries taking place inside one's own mind possibly shed any light on the universe beyond? Why should the material universe obey immaterial laws?

We have seen that the biblical worldview provides the answer: God has given the mind of man some access to His own thoughts – and it is God's thoughts that uphold and control the physical creation. When human beings do mathematics, they are "thinking God's thoughts after Him" and such thoughts will have application to the physical world. So of course the physical creation will obey mathematical laws. And God has made at least some of these laws simple enough to be discoverable by the human mind. It makes sense – in the biblical worldview.

But as Wigner pointed out, the secular worldview cannot make sense of this, and therefore cannot make sense of the success of science. He states, "It is difficult to avoid the impression that a miracle confronts us here, …or to the two miracles of the existence of laws of nature and of the human mind's capacity to divine them." This "miracle" is what makes science possible. And yet it is utterly without explanation in the secular worldview. Wigner ends by concluding, "The miracle of the appropriateness of the language of mathematics for the formulation of the laws of physics is a wonderful gift which we neither understand nor deserve." He is right; from his secular perspective, the success of applying mathematical reasoning to the physical universe is utterly baffling. But from the biblical perspective, it is perfectly reasonable.

This is perhaps revealed most profoundly in the physics discovered by Einstein. The constancy of the round-trip speed of light and the relativity principles are based on observation. But the rest of relativity we were able to *discover and prove by principles of reasoning* without the need for external experiments. And yet all external experiments to date have verified our discoveries. They must do so because the same Mind that governs mathematics also controls all the physical world. Throughout this book, all our discoveries were made possible because we were thinking God's thoughts after Him.

It is an awesome thought. The Lord who controls space, time, and matter in the delightful way we have discovered in this text is the same God who spoke to Abraham and Moses, parted the Red Sea, became man, died our death, and rose again, demonstrating His power of life and death. And despite our treasonous rebellion against Him, He has offered forgiveness and eternal life to those who will repent and trust in Him, confessing Christ as Lord.

A proper study of science not only glorifies God, it humbles man. What mere human could have devised such a remarkable relationship between space, time, and mass, such that each one relates to the other two and yet with no contradictions? The history of science has shown that God upholds the universe in such a way that we can continually discover new truths about it, without ever running out of new mysteries to explore. At the present time, there is a great mystery concerning how to reconcile relativity with quantum mechanics. Each field makes correct predictions in appropriate areas, and yet there seems to be a fundamental incompatibility between them. Perhaps future scientists will discover the answer, but from history we know that this will only open up new mysteries. How could we expect anything less from the mind of God?

Derivation of $E = mc^2$

In chapter 14, we saw that the formula for relativistic mass implies that $E = mc^2$ in cases where the velocity is much less than the speed of light. We analyzed only the low-velocity case so that we could make two approximations; we used the classical kinetic energy formula, and we neglected higher order terms in the binomial expansion of the relativistic mass formula. This allowed us to derive the expression for energy without the use of calculus. Here we show a more sophisticated way to prove that $E=mc^2$ without making any approximations or low-velocity assumptions. We start with the definition of force and energy.

Force is defined by the following formula:

$$F = \frac{d\mathbf{p}}{dt}$$

We define mass to be related to momentum as $\mathbf{p} = m\mathbf{v}$, where m is the relativistic mass. In classical physics (dealing with situations in which $v \ll c$), the mass is essentially identical to the rest mass, which does not change with velocity. And the time derivative of velocity is acceleration. So, we have:

$$F = m\frac{d\mathbf{v}}{dt} = ma \qquad \text{(Classical limit)}$$

This of course is Newton's second law. However, in relativity, mass is no longer constant. Therefore, Newton's second law no longer applies, though it is a good approximation at low speeds.

The definition of kinetic energy is a force (F) applied over a distance (dr) as follows:

$$E = \int_{r_0}^{r_1} F \cdot dr$$

If we consider the scenario in which velocity and force are limited to one dimension, then we have:

$$E = \int_{x_0}^{x_1} F dx$$

In the classical limit where $v \ll c$, we can use the classical approximation for force (Newton's second law) to find the classical formula for kinetic energy:

$$E = \int_{x_0}^{x_1} (ma)dx = \int m\frac{dv}{dt}dx = \int m\frac{dx}{dt}dv = \int_0^v mvdv = \frac{1}{2}mv^2$$

But this derivation doesn't work for high velocities because mass isn't constant, and therefore **F=ma** is no longer true. So, starting with the definitions of energy and force we have:

$$E = \int_{x_0}^{x_1} F dx = \int \frac{dp}{dt}dx = \int d(mv)\frac{dx}{dt} = \int d(mv)v$$

Applying the chain rule to the differential of mv we find:

$$E = \int (mvdv + v^2 dm)$$

To solve this integral, we will need to use the relationship between m and v: the relativistic mass formula.

$$m = \frac{m_0}{\sqrt{1 - \frac{v^2}{c^2}}}$$

Squaring this equation and rearranging terms we have:

$$m^2c^2 - m^2v^2 = m_0^2c^2$$

Taking the differential of this formula gives:

$$2mc^2\,dm - 2m^2v\,dv - 2mv^2\,dm = 0$$

Dividing by $2m$ and rearranging terms yields:

$$mv\,dv + v^2\,dm = c^2\,dm$$

But the left-hand side of the equation is exactly the expression in the integral of the energy formula above. So we substitute this expression into the energy formula as follows:

$$E = \int c^2\,dm$$

This yields the following:

$$\boxed{E = mc^2}$$

This expression is derived under the constraint that E is the kinetic energy of the object, and m is the "extra" mass that is due to such energy. However, we saw in chapter 14 that it applies equally well to the total energy of the object, if we allow m to represent the total (relativistic) mass of the object.

Summary of Numbered Equations

The Galilean velocity addition theorem is an approximation that works well only for speeds (v) much less than the speed of light (c). So $v \ll c$. Observer A moves at speed v relative to observer B. Observer A sees an object moving at speed w in the same direction that he is moving relative to B. Observer B will see the object moving at speed W.

$$W = v + w \qquad (2.1)$$

The time dilation formula relates the time (T') between two successive ticks of a clock (moving at speed v) relative to the time (T) measured by a stationary clock.

$$T = \frac{T'}{\sqrt{1 - \frac{v^2}{c^2}}} \qquad (4.1)$$

The length contraction formula relates the length (L') measured by an observer moving at speed v relative to the length (L) measured by a stationary observer.

$$L = L'\sqrt{1 - \frac{v^2}{c^2}} \qquad (5.1)$$

The Lorentz factor (γ) can be thought of as the "strangeness" factor of relativity. It has a value of 1 for a stationary object, and approaches infinity as an object approaches the speed of light.

$$\gamma = \frac{1}{\sqrt{1 - \dfrac{v^2}{c^2}}} \tag{5.2}$$

The length contraction equation is expressed as a function of the Lorentz factor.

$$L = {L'}/{\gamma} \tag{5.3}$$

The time dilation equation is expressed as a function of the Lorentz factor.

$$T = T'\gamma \tag{5.4}$$

The Galilean transformation takes the coordinates of an event that takes place at x, y, z, t in a stationary frame, and gives the coordinates (x', y', z', t') in a frame moving at speed v in the positive x-direction and overlapping with the stationary frame at $t = 0$. The transformation is an approximation that gives good results only at low speeds ($v \ll c$).

$$\begin{aligned} x' &= x - vt \\ y' &= y \\ z' &= z \\ t' &= t \end{aligned} \tag{8.1}$$

The Lorentz transformation takes the coordinates of an event that takes place at x, y, z, t in a stationary frame, and gives the coordinates (x', y', z', t') in a frame moving at speed v in the positive x-direction and overlapping with the stationary frame at $t = 0$. The transformation is exact and gives correct answers at all speeds.

$$\begin{aligned} x' &= \gamma(x - vt) \\ y' &= y \\ z' &= z \\ t' &= \gamma\left(t - \frac{vx}{c^2}\right) \end{aligned} \tag{8.2}$$

The inverse Lorentz transformation takes the coordinates from the moving frame and expresses them in terms of coordinates in the stationary frame:

$$x = \gamma(x' + vt') \tag{8.3}$$
$$y = y'$$
$$z = z'$$
$$t = \gamma\left(t' + \frac{vx'}{c^2}\right)$$

The relativistic velocity addition theorem gives the velocity (W) of an object relative to a stationary frame when the object is measured to have velocity w relative to an observer moving at speed v in the same direction. The formula is exact, works at all velocities, and is the relativistic replacement for the Galilean velocity addition theorem.

$$W = \frac{v + w}{1 + \frac{vw}{c^2}} \tag{8.4}$$

Useful in geometry, the rotation formulae give the coordinates of a point in one coordinate system relative to those in another coordinate system in the same plane with the same origin but rotated by angle θ.

$$x' = x\cos(\theta) + y\sin(\theta) \tag{8.5}$$
$$y' = y\cos(\theta) - x\sin(\theta)$$

The difference between two values is expressed by an uppercase delta (Δ). So, the difference in values between two events in spacetime is expressed as follows:

$$\Delta x = x_1 - x_0 \tag{9.1}$$
$$\Delta y = y_1 - y_0$$
$$\Delta z = z_1 - z_0$$
$$\Delta t = t_1 - t_0 \tag{9.2}$$

The distance formula gives the distance (r) between two points in space.

$$r = \sqrt{\Delta x^2 + \Delta y^2 + \Delta z^2} \tag{9.3}$$

The spacetime interval (s) is analogous to the distance between two points in space, but gives the "distance" in both space and time between two events:

$$s = \sqrt{\Delta x^2 + \Delta y^2 + \Delta z^2 - c^2 \Delta t^2} \qquad (9.4)$$

To avoid imaginary numbers, the spacetime interval is often expressed by its square:

$$s^2 = \Delta x^2 + \Delta y^2 + \Delta z^2 - c^2 \Delta t^2 \qquad (9.5)$$

Some physicists prefer to define the spacetime interval by the opposite sign. Using this alternate definition, the square of the spacetime interval is given as follows:

$$s^2 = -\Delta x^2 - \Delta y^2 - \Delta z^2 + c^2 \Delta t^2 \qquad (9.6)$$

Proper time (τ) is the time difference between two events measured by an observer who is at the location of each event when it occurs.

$$\tau^2 = \Delta t^2 - \frac{\Delta x^2}{c^2} - \frac{\Delta y^2}{c^2} - \frac{\Delta z^2}{c^2} \qquad (9.7)$$

The square of proper time is essentially the negative square of the spacetime interval, measured in units of time rather than units of distance. The two are related by the following formula:

$$\tau^2 = -\frac{s^2}{c^2} \qquad (9.8)$$

The relationship between frequency (ν) (this is the Greek letter "nu" and not the letter "v") and wavelength (λ) of light traveling in vacuum.

$$\lambda \nu = c \qquad (11.1)$$

The classical Doppler shift formula gives the observed wavelength of light (λ_o) relative to the source wavelength (λ_s) when the source is receding at velocity v. This formula is an approximation that gives accurate results only when the source is moving much slower than the speed of light.

$$\lambda_O = \lambda_S \left(1 + \frac{v}{c}\right) \tag{11.2}$$

The relativistic Doppler shift formula includes the effects of time dilation and length contraction. It gives the observed wavelength of light (λ_O) relative to the source wavelength (λ_S) when the source is receding directly away from the observer at speed v. The formula is exact and gives accurate results at all velocities.

$$\lambda_O = \lambda_S \sqrt{\frac{1 + \frac{v}{c}}{1 - \frac{v}{c}}} \tag{11.3}$$

The transverse Doppler shift formula gives the observed wavelength of light (λ_O) relative to the source wavelength (λ_S) when the source is moving *perpendicular* to the observer at speed v.

$$\lambda_O = \gamma \lambda_S \tag{11.4}$$

The formula for redshift due to acceleration (a) between two points separated by distance h is given below. The formula is an approximation giving good results when $ah \ll c^2$.

$$\lambda_O = \lambda_S \left(1 + \frac{ah}{c^2}\right) \tag{11.5}$$

The formula for time dilation due to acceleration is given below. The observer who is behind (A) will experience greater time dilation (his clocks will tick slower) than the observer who is ahead (B), even though both move at the same speed and accelerate at the same rate. The formula is an approximation for cases where $ah \ll c^2$.

$$\frac{\Delta T_B}{\Delta T_A} = 1 + \frac{ah}{c^2} \tag{11.6}$$

The formula for Lorentz aberration where $\theta = 90°$ gives the angle at which an object will appear relative to a moving observer if the object's position is perpendicular to the direction of motion as seen by the stationary observer. For example, a star that is directly overhead ($\theta = 90°$) for observer A, will appear at angle θ' for observer B, if observer B is moving at high speed along the ground.

$$\tan \theta' = \frac{c}{\gamma v} \qquad (12.1)$$

The formula for Lorentz aberration gives the angle (θ') that a moving observer sees for an object if the stationary observer sees the object at angle θ.

$$\tan \theta' = \frac{\sin \theta}{\gamma \left(\cos \theta + \frac{v}{c} \right)} \qquad (12.2)$$

The formula relating the momentum (**p**) of an object to its mass (*m*) and velocity (**v**) is given below. The formula is exact if *m* is defined as relativistic mass.

$$\boldsymbol{p} = m\boldsymbol{v} \qquad (13.1)$$

A change in momentum ($\Delta \boldsymbol{p}$) is produced by an unbalanced force (**F**) over a time interval (Δt).

$$\Delta \boldsymbol{p} = \boldsymbol{F} \Delta t \qquad (13.2)$$

In one dimension, the change in kinetic energy (ΔE) of an object is the net force (*F*) applied on the object multiplied by the change in position (Δx):

$$\Delta E = F \Delta x \qquad (13.3)$$

The relativistic mass formula gives the relativistic mass (*m*) of a moving object relative to its rest mass (m_0), often expressed using the Lorentz factor (γ).

$$m = \gamma m_0 \qquad (13.4)$$

The momentum (p) of an object moving at velocity v expressed in terms of its rest mass (m_0):

$$p = \gamma m_0 v \tag{13.5}$$

The classical formula for kinetic energy (E) is an approximation that gives accurate results only in cases where $v \ll c$.

$$E = \frac{1}{2}m_0 v^2 \tag{14.1}$$

The relativistic energy formula gives the total energy (E) of an object in terms of its relativistic mass (m) and is exact at all velocities.

$$E = mc^2 \tag{14.2}$$

The relationship between total energy (E), rest mass (m_0) and momentum (p) is given below.   The formula is exact and works at all velocities.

$$E^2 = (m_0 c^2)^2 + (pc)^2 \tag{14.3}$$

The energy (E) of a photon in terms of its momentum (p) is given by:

$$E = pc \tag{14.4}$$

The total energy (E) of an object in terms of its rest mass (m_0) is given by:

$$E = \gamma m_0 c^2 \tag{14.5}$$

Newton's law of gravity shows the gravitational force between to masses (m, and M) separated by a distance (r). G is the universal gravitational constant and has a value of 6.67408×10^{-11} $m^3 kg^{-1} s^{-1}$.

$$F = \frac{-GMm}{r^2} \tag{15.1}$$

Newton's second law gives the relationship between a net force (F) on an object, the object's mass (m), and its acceleration (a). The law is an approximation and works well only for velocities much less than the speed of light.

$$F = ma \qquad (15.2)$$

The acceleration of gravity of an object in vacuum toward a mass (M) at distance r.

$$a = \frac{-GM}{r^2} \qquad (15.3)$$

The formula for the *difference* in gravitational time dilation between two observers whose elevation difference is h is given below. Here the acceleration of gravity is g (for sea level on Earth $g = 9.8$ m/s²). The observer at higher elevation (B) will experience less gravitational time dilation than the other observer (A). The equation is an approximation for cases where $gh \ll c^2$.

$$\frac{\Delta T_B}{\Delta T_A} = 1 + \frac{gh}{c^2} \qquad (15.4)$$

Escape velocity (v_E) is the speed at which an object has sufficient kinetic energy to escape the local gravitational field. The source of gravity has mass M and is at a distance r. The universal gravitational constant is G.

$$v_E = \sqrt{\frac{2GM}{r}} \qquad (15.5)$$

The formula for gravitation time dilation gives the slowing of clocks in a gravitational well relative to an observer at infinite distance. This formula is more exact than equation 15.4, and applies to any spherical source of gravity of mass M, where the object is located outside the sphere at distance r from the center.

$$\frac{\Delta T_B}{\Delta T_A} = \frac{1}{\sqrt{1 - \frac{2GM}{rc^2}}} \qquad (15.6)$$

The gravitational redshift formula can be used to compute the wavelength of light as it escapes a gravitational well. Here λ_O denotes the observed wavelength of light, λ_S denotes the source wavelength, r is the distance of the source from the center of the spherical mass, and M is the mass.

$$\frac{\lambda_O}{\lambda_S} = \frac{1}{\sqrt{1 - \frac{2GM}{rc^2}}} \tag{15.7}$$

The Schwarzschild radius (R_S) is the "point of no return" for a black hole of mass M.

$$R_S = \frac{2GM}{c^2} \tag{16.1}$$

The tidal force formula gives the difference in the force of gravity (ΔF) exerted on two sides (separated by Δr) of an object of mass m at a distance r from the source of gravity of mass M.

$$\Delta F = \frac{2GMm}{r^3} \Delta r \tag{16.2}$$

The epsilon time dilation formula compares the passage of time between two reference frames leaving the choice of synchrony convention as a free parameter (ε). Under the standard synchrony convention, $\varepsilon = \frac{1}{2}$ and the equation 17.1 reduces to equation 5.4. Under the anisotropic synchrony convention, $\varepsilon = 1$.

$$T' = T\sqrt{1 - \frac{v^2}{c^2}} \cdot \frac{c}{c + v(2\varepsilon - 1)} \tag{17.1}$$

The epsilon length contraction formula compares the length of a moving ruler with that of its rest length leaving the choice of synchrony convention as a free parameter (ε).

$$L = L' \sqrt{1 - \frac{v^2}{c^2}} \cdot \frac{c}{c + v(2\varepsilon - 1)} \qquad (17.2)$$

The formula for the one-way speed of light (c_θ) under the anisotropic synchrony convention (ASC) depends on the angle (θ) of the direction of light relative to that of a directly incoming beam. Here, c denotes the round-trip speed of light.

$$c_\theta = \frac{c}{1 - \cos \theta} \qquad (18.1)$$

The following formulae convert the coordinates of an event in spacetime from the standard convention (x, y, z, t) to the anisotropic synchrony convention $(x_\varepsilon, y_\varepsilon, z_\varepsilon, t_\varepsilon)$.

$$x_\varepsilon = x \qquad (18.2)$$
$$y_\varepsilon = y$$
$$z_\varepsilon = z$$
$$t_\varepsilon = t + \frac{r}{c}$$

Where

$$r = \sqrt{x^2 + y^2 + z^2}$$

Made in the USA
Middletown, DE
01 May 2019